3-15

# TO PUNISH OR PERSUADE

# TO PUNISH OR PERSUADE

*Enforcement of Coal Mine Safety*

JOHN BRAITHWAITE

State University of New York Press
Albany

Published by
State University of New York Press, Albany

© 1985 State University of New York

Printed in the United States of America

No part of this book may be used or reproduced
in any manner whatsoever without written permission
except in the case of brief quotations embodied in
critical articles and reviews.

For information, address State University of New York
Press, State University Plaza, Albany, N.Y., 12246

**Library of Congress Cataloging in Publication Data**

Braithwaite, John.
    To punish or persuade.

    1. Coal mines and mining—Accidents—Case studies.
2. Coal mines and mining—Accidents—Government policy.
3. Mine safety—United States—Case studies.  I. Title.
TN311.B68  1984     363.1'19622334     84-2671
ISBN 0-87395-931-0
ISBN 0-87395-932-9 (pbk.)

*My lovely boys, my poor tender boys;*
*oh, where are they?*

—A mother near tunnel entrance,
at Bulli, 1887, where 81 died

# Contents

# Preface

I was brought up in Ipswich, a coal-mining town in Queensland, Australia. Two men with whom I once played cricket and football have been killed in mine accidents. But the real motivation for writing this book was the Box Flat disaster of 1972—all of us in Ipswich were profoundly moved and disturbed when seventeen men were entombed near our homes.

After Box Flat, I expected one day to write an angry book about death in the mines. This modest and dispassionate contribution hardly meets that aspiration, but I have come to the conclusion that it might ultimately benefit miners more than a book directing fire and brimstone at those responsible for mine fatalities.

I owe a debt of gratitude to the many miners, company executives, union officers, inspectors, and other government officials who cooperated in the research. It is unfair to single people out, but during my two periods of fieldwork in the United States, Ron Schell, John Greenhalgh, and Davitt McAteer were especially helpful.

I am grateful for the support of the Center for the Study of Values, at the University of Delaware, where I was a Visiting Fellow during late 1981 and early 1982, and to my employer during most of this research project, the Australian Institute of Criminology.

PLAN OF THE BOOK

This book is more an attempt at policy analysis rather than social science. Consequently it contains some inferential leaps that will leave social scientists uncomfortable. This is inevitable

ix

in any book seeking to advocate the direction of public policy that is most supportable in light of our currently inadequate empirical understanding of the problems.

The goal of this policy analysis is to find the method of punishment (defined broadly as any formal or informal sanction imposed for violations of law) that will achieve the greatest reduction of carnage in coal mines.

The book has two parts. Part I contains the empirical background necessary for an understanding of the organizational and institutional (as opposed to technical and geological) forces responsible for coal mine accidents. The core consists of studies of the causes of thirty-nine mine disasters since 1960 in five countries (chapter 2) and an analysis of the safety compliance systems of the five current corporate leaders in coal mine safety in the United States (chapter 3).

Part II addresses the policy question of when, if ever, to use punishment as a regulatory strategy and when to use persuasion. Persuasion means attempts to change safety practices by advice, education, and entreaty. Chapter 4 considers, in turn, whether punishment works as a regulatory strategy, whether persuasion works, and whether punishment and persuasion are incompatible strategies. The subsequent chapter examines tactics for bringing punishment to bear in such a way as to minimize its interference with persuasion and to maximize synergy between the two approaches. Chapter 6 discusses how punishment is best administered—who should do it, against what targets, and using what kinds of sanctions under what kind of structure of regulatory decision making. In chapter 7, consideration is given to whether the impact of law enforcement on productivity should be incorporated into the policy analysis. Chapter 8 pulls together some of the broader conclusions.

The policy implications are general and international, rather than fixed on how to reform a particular law in a particular country. Research for the book took me to the United States, Great Britain, Australia, France, Belgium, and Japan, though the emphasis is on data from the first three countries. If no source for information in the text is cited, it is interview data or unpublished material supplied by the regulatory agency of the country concerned, or by the management of the company involved. Readers from other countries may be mystified by my discussion of Australia's two major coal-mining states—New South Wales and Queensland—as if they are separate countries. With respect to mine safety and health laws, they are: there are no national laws, only state laws.

Male personal pronouns have been used to refer to mining employees. The fact that I come from a state where it is still illegal for women to be employed underground does not excuse the sexist language; it merely explains it.

# Part I

# The Empirical Background

# 1. Introduction

THE PROBLEM

The toll of death in the world's coal mines has been horrifying. Some 60,000 miners have been recorded as killed in British coal mine accidents this century. In the United States the toll is over 100,000. These figures represent only the deaths from accidents—the toll from occupational disease, particularly black lung disease, has been even greater. A 1965 survey found that, conservatively, there were 100,000 retired coal miners in the United States suffering gasping breathlessness from black lung (Franklin 1969: 122). One study found that, even *excluding* the probability of death from accidents or violence, coal miners had death rates ranging from 23 percent higher than the national male worker's average for 20-24-year-olds to 122 percent higher for 60-64-year-olds (Enterline 1964: 761).

While the loss of life is frightening, the improvement has been encouraging. In the United States the death toll has declined from a peak of 3,241 in 1907, to 70 in 1983; in Britain it has dropped from 1,484 in 1866, to 44 in 1982-83. Japanese statistics do not go back so far, but Japanese coal mine fatalities have dropped from 347 in 1966, to 22 in 1980. (However, this increased again in 1981, when a single disaster claimed 93 lives, and in 1984 when another 83 perished in one fire.) Such improvement is evident in every major coal-mining country. Massive loss of life in single disasters was commonplace early in the century. One French disaster in 1906 claimed 1,099 lives. Wholesale slaughter still occasionally occurs in countries with nineteenth-century safety laws such as South Africa and Rhodesia (now Zimbabwe) which, as recently as 1973, had a disaster in which 427 miners perished. Today some third world countries, such as India, have more fatalities than the world leader in coal production, the United States, and even countries with relatively minor coal industries, such as South Korea, are averaging 100 fatalities a year in their coal mines.

Nevertheless, to look only at fatalities is grossly to understate the size of the problem. For every fatality, there are hundreds of injuries to coal miners serious enough to cause time off work. Australia's largest state, New South Wales, does not have a staggering death toll; however, each year there are more than 10,000 compensable injuries, a remarkable number when one considers there are only 15,000 miners in the New South Wales coal industry. In spite of the small number of coal miners in the state, estimates from Australia's National Crime Victims Survey suggest that the number of occasions when New South Wales citizens require medical treatment as a result of assaults or robberies is lower than the annual number of injuries to coal miners requiring medical treatment. The $26 million paid in compensation to injured miners in 1978-9 represented 9.7 percent of the gross wages paid for the period.

A conclusion of this book is that enforcement of mine safety laws has been a major factor in producing the dramatic improvements in coal mine safety of the past century, and that tougher enforcement in future can produce further improvement. This is not to say, the tougher the enforcement the better. There are ways in which a prosecutorial approach can hinder rather than help coal mine safety. Sometimes, as in all areas of law enforcement, the inspector will better serve the people he seeks to protect by persuasive or educative appeals to the better nature of offenders than by prosecution. The mission of this book is to begin to define an optimal mix of punishment and persuasion. In delineating a strategy for combining punishment and persuasion that attempts to maximize the protection of coal miners, a model is developed that may be applicable to many areas of business regulation.

The analysis is unreservedly utilitarian. The concern is to find a level of punishment for violators of coal mine safety laws that provides the maximum protection for workers. To sacrifice the lives of miners for the sake of giving offenders their "just deserts" is morally unacceptable to this author. Readers of a retributive bent should be prepared for an argument for a lower level of punishment for offenders who cause death and injury to workers than most members of the community would perceive as just. A growing body of evidence suggests that in many countries the community has a surprisingly punitive attitude toward corporate offenses that cause loss of life or serious injury to persons (Braithwaite 1982a). It follows from the argument in this book that, if the community wants to indulge these sentiments fully, it will do so at the expense of further loss of life and limb. It will also be argued, however, that a far more punitive approach than presently exists in any country is needed to reduce mine fatalities further.

No society has ever adopted a truly punitive approach to the regulation of coal mine safety. The most aggressive mine safety enforcement regime existed under the Carter Administration in the United States. In 1980, some 140,000 fines were imposed on coal-mining companies for health and safety violations in the United States. However, the fines were small, averaging $173 after reductions on appeal. Only four persons between 1978 and 1984 were sentenced to periods of incarceration for coal mine health and safety offenses. Following the 1968 Farmington disaster, in which seventy-eight miners perished, the U.S. enacted probably the world's most punitive statute for regulating business, the 1969 Federal Mine Safety and Health Act. Much more is said on this legislation in later chapters. Even though the Reagan Administration has substantially toned down the punitiveness of the legislation in practice, the broader historical trend of coal mine safety enforcement in the United States has been toward increasing punitiveness. New South Wales and Japan have also become slightly more punitive in the last few years as a result of major disasters. However, at least for Australia, this slight increase during the early 1980s must be balanced against a reduction in punitiveness over the greater part of this century.

Great Britain displays a striking pattern of decreasing punitiveness; the average number of prosecutions of owners and managers per million employees per annum dropped from 27 for 1911-20, to 15 for the 1920s, 9 for the 1930s, 5 for the 1940s, and 1 or less in the decades since nationalization (Collinson 1976; 3.6).

To illustrate the shift more dramatically, in the years immediately prior to World War I, an average of 32 mine owners,
agents, managers, and undermanagers were prosecuted for an
average of 157 offenses per annum, resulting in an average of 89
convictions; the average annual charges against lesser officials
and workmen was 1,152 (Royal Commission on Safety in Coal
Mines 1938, pp. 84, 92). In 1980 and 1981, there were no
prosecutions at all in Great Britain for mine safety violations. In
France, prosecutions have also fallen away to nothing since
nationalization of the industry after World War II. The last
prosecution there seems to have followed a disaster in 1974.
Cynics might be right in interpreting postnationalization
nonpunitiveness of mine safety enforcement in Britain and
France as evidence of the hopelessness of a government's
regulating itself, but it must also be viewed as part of a much
longer-term shift away from prosecution and toward persuasion
as the best way of achieving mine safety.

Hence, on opposite sides of the Atlantic we have, on the one
side, a strong historical shift away from punitiveness in coal
mine safety enforcement and, on the other side, a movement
toward punitiveness. This book will not examine the historical
forces that have produced such a dramatic divergence in societies
which normally follow rather similar trends; instead, we will be
limited to an analysis of the arguments that regulators on both
sides advance to sustain the proposition that it is they who are
following the right course. Almost all countries have devoted
increasing resources to the problem over the past decade. The
U.S. Mine Safety and Health Administration now has some
1,400 inspection personnel, but it is the Japanese with 110
government inspectors—half of them university graduates—for
their thirty-two mines, who have shown the greatest proportionate investment in coal mine safety supervision. In addition
to establishing the justification for this kind of investment, this
book, it is hoped, will provide some ideas on how these precious
resources might be better utilized.

## OTHER APPROACHES TO IMPROVING COAL MINE SAFETY

Although it will be argued that more effective enforcement
of better laws has played a substantial role in improving the

safety of coal mines, it must be said at the outset that this is only one of a number of important considerations. Sir Andrew Bryan has listed such additional factors:

> ...the general advance in knowledge of the causes of mining dangers; the invention and adoption of new safeguards and safety devices; the increasing use of protective clothing; the growth in mechanisation, remote control and automation; the greater use of specialist engineers and scientists; the growth of planning; the development of the concept of in-built safety in mining methods and machines; the continuing improvement in the facilities and opportunities for the technical education of officials and work-men; the growth of "on-the-job" training; the gradual increase over the years not only in the number but also in the competence of full-time workers in the field of accident prevention who now included a higher proportion of university graduates; and the intensification in recent years of safety campaigns... (Bryan 1975; p. 96).

In Europe these factors have combined with increased inspections and better regulations to improve safety even though natural conditions (thinner and deeper seams) have become more hazardous. Mechanization has been the most fundamental factor in reducing the number of deaths, because it has meant that fewer miners are needed and threrefore fewer are exposed to danger. On the other hand, it has been argued that one important technological advance, the continuous miner, has worsened dangers for those remaining in the mine (David 1972). Since the principal advantage of the continuous miner is the high speed with which it rips coal from the face, effectively burrowing itself into the seam, the operator constantly exposes new, unsupported roof. The more rapid rate of advance increases the making of explosive methane gas and throws up more dust, thereby worsening exposure to pneumoconiosis (black lung disease). The machinery has also made hearing loss a severe health problem for coal miners. The fast movement of the machines around poorly lit areas has also been a new source of accidents, while the electrical cables that trail behind them create new ignition sources for explosions. However, with time, new regulations have been devised—water sprays to suppress dust, automatic temporary roof-support systems, improved illumination in areas where machines work—and these are beginning to control the unsafe aspects of the new technology to such a point that today it is indisputable that mechanization has on balance reduced the loss of life and limb in our mines.

The potential for technology further to lower death and injury in coal mines is enormous. The British National Coal Board's chief safety engineer has analyzed all fatal accidents in Britain between 1975 and 1980 and concluded that 43 percent of them could have been prevented by the application of existing and affordable accident prevention technology (Collinson 1980; 1976; 1979). This raises a question not addressed in this book: Would more lives be saved if governments of countries with large coal-mining industries redeployed scarce mine safety resources from enforcement to investment in safety technology research and development? This is certainly a question deserving of greater attention. However, it sets up a rather artificial distinction between expenditure on technological innovation and enforcement. Much enforcement effort is directed at mandating new, improved safety technology, or at setting performance standards (e.g., an acceptable ambient dust concentration in a mine environment), which force industry to innovate new technologies to meet the standards. Equally, moreover, one can ask whether more of the research and development budgets of governments should be diverted from new technology that improves productivity to technology that improves safety. A National Academy of Sciences (1969, p. 82) study found that, in 1969, only 5 percent of the U.S. Bureau of Mines reserach budget for new technology development was being devoted to health and safety projects. The situation may be better today—the bureau spent $86 million on health and safety research in 1982; research on technology to improve productivity has been transferred to the Department of Energy.

Improved training can also considerably reduce mining accidents; the data in this book affirms this. The conventional wisdom of the industry is that a large part of the reason for Britain's having the safest mines among the world's major coal producers is that it has the most demanding training requirements for miners and certification requirements for managers. The resources deployed are also impressive.

> The NCB [National Coal Board] employs 1,200 full-time training
> staff members to work with the 250,000 British coal mines at forty
> training centers throughout the country. Each mine is assigned at
> least one full-time training officer, who is responsible for the
> miner training program. Training for managerial personnel is
> conducted at two staff colleges... (McAteer and Galloway 1980,
> p. 944).

Within the British system there is evidence to suggest that
the investment in training produces results. Until recently, fifty
pits were being selected each year for a special safety campaign
during the year. These pits consistently showed a greater
improvement in accident rates than other British pits (Collinson
1978, p. 77).

Industry lobby groups are forever urging a shift of government priority from enforcement to training. But this, too, may
create a false dichotomy. Today many regulations relate to
requirements that mines have training programs of various
kinds and that persons not be permitted to perform managerial
or technical tasks unless they have satisfactorily completed
certain training. Moreover, most of the investment in training
has always been, and will always be, by mine operators rather
than regulatory agencies. So, the more fundamental way of
improving training is to effect a reallocation of mine operators'
budgets rather than a realignment of regulatory budgets. Again,
however, I would not want to deny that more resources and
research ought to be directed at reducing accidents by improved
training.

Beyond training and technology, if we are to have a broad
understanding of why mines have become safer over the last
century, institutional factors must be considered as well. Two
are fundamental: the growth of the power and safety consciousness of trade unions and the postwar tendency for less socially
responsible corporations to be swallowed up by more responsible ones.

Particularly in the United States, trade unions were not
always vanguards of safety activism. Rather, they were often
captives of the productive interests of capital. In 1947, the United
Mine Workers (UMW) in Illinois voluntarily signed a labor
contract with terms forbidding the union from seeking im-

provements in Illinois mine safety law, upon which the industry placed so much store in resisting federal control (Franklin 1969, p. 127). However, since the change of leadership, in 1972, the UMW, which still dominates the American coal work force, has become a more activist union in the cause of occupational health and safety. The proportion of wildcat strikes accounted for by safety issues increased from 8 percent in 1970 to 13 percent in 1974 (General Accounting Office 1981, p. 20). For some years, contracts between the UMW and the Bituminous Coal Owners Association have given considerable powers to Mine Safety and Health Committees of miners, selected by the local union. The 1981 contract provides, inter alia, that:

> In those special instances where the Committee believes that an imminent danger exists and the Committee recommends that the Employer remove all Employees from the involved area, the Employer is required to follow the Committee's recommendation and remove the employees from the involved area immediately (p. 11).

The UMW also has a safety department with fifty-eight field inspectors. These are obviously insufficient to provide any sort of coverage of all the mines in the United States. However, they are sufficient to swoop down on mines where the management has been reported by the rank and file to be flouting the recommendations of local Health and Safety Committees. Managers who have undermined their Health and Safety Committees have been known to endure full-time union safety inspectors camped at their mine until literally hundreds of violations have been recorded and reported to the Mine Safety and Health Administration. Years before the law imposed mandatory training requirements on mine operators, UMW contracts did so.

Perhaps as a result of these activities, fatality statistics support the conclusion that UMW members are less likely to be killed in American mines than are nonunionists. One reason may be that nonunionized mines are more likely to be small mines (National Research Council 1982, p. 12); but equally, one reason small mines have high fatality rates might be the unwillingness of miners to disobey instructions from small mine

managers to follow unsafe practices when they know there is no union to back them up. The *UMW Journal* ran a front-page story in its November 1981 issue, pointing out that only 33 percent of the eighty-nine miners killed so far that year were UMW members, while 70 percent of the nation's miners were UMW members. Even though this comparison probably exaggerated the difference because the period included a UMW strike, there can be little doubt that nonunionists in the United States suffer higher fatality risks. Of the UMW's 2,300 mines, 1,900 have accident rates below the national average. The union is running a campaign of safety seminars and inspections targeted at the minority of unionized mines with accident rates above the national average. The goal is to bring every union mine below the average. As of December 1981, 69 union mines having above average accidents had been brought below the average.
brought below the average.

On the other hand, there is some evidence from U.S. studies that unionized mines can actually have higher nonfatal accident rates than nonunionized mines (Boden 1983; DeMichiei et al. 1982, p. 27). However, this data trend is also consistent with the interpretation that, at UMW mines, the union monitors the reporting of accidents to ensure against underreporting (Boden 1983). Other more qualitative evidence does suggest that the UMW can and does have a positive influence on nonfatal accidents. A comparison of high-accident and low-accident mines by Pfeifer, Stefanski, and Grether (1976, p. 115) found that both safety directors and miners in mines with low accident rates reported that the union put greater pressure on management for safer and more healthful mines through bargaining and discussing safety and health topics in union meetings than was reported by safety directors and miners at mines with high accident rates.

In many countries around the world, coal-mining unions are unique for their solidarity and for the concessions this has won them in the health and safety of their mines. For a hundred years British miners have had the right to appoint their own inspectors to check, without interference from employers, the safety of workplaces. Other British workers have won this right only in the last decade. Countries as diverse as Australia, Japan, the United States, France, Romania, and Poland have followed the British lead in conferring rights of inspection on miners' representatives. Whatever we think of the realities of union power in Poland, the law does formally concede miners' representatives the authority to order management to abate violations and order closure of mines, or stoppage of work at the face, when regulations are not being followed (McAteer and Galloway 1980, p. 972). Union check inspectors have the same power in Queensland, Australia. In many countries, coal-mining unions are among the few that employ a body of full-time workers' inspectors to check the safety of work sites.

The second major institutional force has been the tendency since World War II for less safety-conscious mining companies to be taken over by larger, more socially responsible corporations. The leading example is the nationalization of the British coal industry, in 1946. This paved the way for Britain's having the safest coal-mining industry in the world, even though their geological conditions are objectively more dangerous than in many other countries, including the United States. We have already mentioned that British fatality rates per worker-hour are about a third of those of the United States, Germany, or Australia, and that this difference may in considerable measure be attributable to Britain's superior safety training. It was in the wake of nationalization that Britain assumed its position of international leadership in quantity and quality of training. The other fundamental effect of nationalization was a progressive raising of the safety standards of those mines with the most poorly managed safety programs toward the standards of the industry leaders. Standards that were acceptable to a minority of private mine owners were intolerable to the National Coal Board (NCB). The centralized direction the industry acquired with nationalization also meant that as soon as a new hazard was identified following a major accident, instructions went out to all NCB mines to introduce a method of controlling the hazard. This is only a more widespread manifestation of what happens with large, reputable private companies. If U.S. Steel has a disaster in one of its mines, and this identifies a new hazard, in no time all U.S. Steel mines will be implementing some measure to prevent a recurrence.

Turton (1981) has plotted the incidence of major explosions in Britain since 1851 and concluded that nationalization was one of two major watersheds that reduced their incidence. Nationalization coincided with a sharp drop in the incidence of explosions taking ten or more lives from an average of one a year to one every five years. Even the unlikely source of a report by the Reagan Administration's Controller General concedes that public ownership might have something to do with Britain's superior coal mine safety performance: "British coal mines are probably the safest because they are managed by the National Coal Board, an arm of the British government." (General Accounting Office 1981, p. 103). It might also be pointed out that the dramatic improvement in safety since nationalization has coincided with a threefold increase in productivity (tons per man shift).

There is also considerable evidence that nationalization of the coal industry in France after the Second World War precipitated a sharp improvement in safety. During the 1940s, 2,674 men were killed in French coal mines. This dropped to 1,753 in the 1950s. The drop is actually less dramatic if you look at the number killed per 10,000 miners, but more dramatic when you consider fatalities per million tons (Dardalhon 1964). The plot of *accident rate* (including nonfatals) per 10,000 miners is actually an inverted U, with accident rates steeply increasing until they reached their peak between 1945 and 1950 and then consistently and steeply decreasing after nationalization (Dardalhon 1964, p. 52).

It is possible, of course, that nationalization has been associated with improved safety performance because of traditional socialist expectations that socialist enterprises will allow safety to interfere with productivity. This may be so, yet one cannot but have the feeling after interviewing executives of these nationalized industries that they are as obsessed with improving productivity as their counterparts in private enterprise. I suspect that the improvements associated with nationalization have had more to do with pulling the standards of the worst mines into line with expectations set by the larger, better managed mines, and the advantages of centralized direction and control for avoiding "carbon copy" accidents.

This suspicion becomes more plausible when one observes similar phenomena in the behavior of large private corporations. British mines might have lower fatality rates than the average U.S. mine. However, the American companies with the best safety records—companies like Old Ben, U.S. Steel, and Bethlehem Steel—have fatality rates in their mines that are about the same, or lower than, those of British National Coal Board Mines. In chapter 3, we look at what these leaders of American coal-mining safety do, and it is fundamentally similar to what the British National Coal Board does.

Four of the American companies leading in safety performance are among the very largest American coal-mining companies. Not all large companies are above average on safety, but there is a clear tendency for the safety leaders to be giant corporations rather than to number among the 92 percent of American coal corporations with assets of $5 million or less (National Research Council 1982, p. 64). The one company among our five safety leaders described in chapter 3 which is only medium in size, Old Ben, underwent the dramatic improvement in accident rate that made it an industry leader after it was taken over by Standard Oil of Ohio, when higher safety expectations were imposed from corporate headquarters. At the other extreme, the least safe sector of the American coal industry is the small mines employing fewer than 50 miners, which are generally owned by little local companies or families (Boden 1983). The fatality rate for mines employing 50 or fewer miners in 1979 was four and a half times as high as for mines employing more than 150. These small mines generally lack a specialist safety and training officer; they do not enjoy the expert advice of accident prevention officers and engineers, who are made available by corporate headquarters; they only infrequently receive instructions on how to avoid the accidents other mines have had; and they do not benefit from safety audits by teams from headquarters who can tell them how other mines are operated more safely.

As in all sectors of the economy, the coal-mining industry is becoming progressively concentrated in fewer and fewer corporate hands. Not all of these emerging giants of coal are paragons of corporate virtue and safety. However, in aggregate, the effect of growing concentration has been to improve health and safety, because more large companies than small ones have the resources and expertise to make mines safer, and also more of them are concerned about the consequences for their diverse operations if they soil their reputation as a responsible company.

If this analysis is correct, it raises some interesting questions about narrowly conceived antitrust laws that permit only the impact on competition to be considered in deciding whether mergers should be allowed. One might suggest that price competition in coal is a trivial consideration compared with saving the lives of coal miners. On this basis, the takeover of a small and unsafe coal-mining company by U.S. Steel would be something to be encouraged, not discouraged. It makes a good case against the narrowly anticompetitive legalism of U.S. antitrust law and for the broad discretion of the British model of a Monopolies and Mergers Commission, which is able to weigh all elements of the "public interest" before making a recommendation to the Secretary of State to issue an order regarding a monopoly or merger. Luckily for coal miners, American antitrust law has been so impotent in achieving its narrowly conceived aims that it has provided little obstacle to the progress of corporate concentration.

## ENFORCEMENT IN PERSPECTIVE

The purpose of the foregoing paragraphs has been to show that improving regulation is not the only route to safer coal mines. And it is not necessarily the best route; it is probably easier to straighten out an unsafe mine by having it taken over by U.S. Steel than it would be by adding government inspectors. Certainly, the limited focus of this book on how best to make enforcement work does neglect important questions about whether we can achieve more by directing our efforts to completely different approaches. Nevertheless, we have at least been able to argue that it would be silly to see incompatibility between enforcement efforts and the other approaches discussed above, because so much of enforcement is directed at requiring or motivating new technologies, mandating training and certification, and guaranteeing that workers can exercise their rights to inspect mines, register complaints, inspect new roof control and ventilation plans, and refuse to work in dangerous conditions.

There may be some who see no place at all for enforcement of coal mine safety standards because they regard collective bargaining by unions for safety (Bacow 1980), market forces informed by worker demands for hazard pay (Viscusi 1983), injury taxes (Smith 1974), or civil litigation to compensate injuries (Posner 1977, pp. 276-8) as providing sufficient incentives for safer mines. However, even as extreme an advocate of unregulated markets as Viscusi (1983, pp. 157-62) sees a limited role for government command and control regulation. Viscusi (1983, p. 160) concedes that it is unrealistic to expect worker demands for hazard pay to induce employer investment in safety (to avoid such hazard pay) when dealing with health hazards that are dimly understood or hazards about which workers cannot get adequate or workplace-specific information. Similarly, strong advocates of collective bargaining, injury taxes, and civil damages litigation tend to concede some place for safety law enforcement, particularly where information costs are insurmountable or where there is inordinate delay in the onset of a health problem. To the extent that this role conceded to enforcement is small rather than large, these theorists will not find the deliberations of this book very central. Nevertheless, for all but the advocates of total safety deregulation, and I have yet to meet one, the book has something to offer.

In the next two chapters, before embarking on a policy analysis of when and how to punish offenders against mine safety and health laws, two empirical studies are undertaken of the causes of disasters and of the characteristics of companies with low accident rates.

# 2. The Disasters Study

The U.S. Mine Safety and Health Administration defines a disaster as an incident in which five or more persons lose their lives. This definition has been adopted here.

Only a minority of coal-mining deaths result from disasters. Most fatalities are caused by such localized incidents as roof falls, or transport or machinery accidents that claim only one or two lives at a time. There is one reason, however, for focusing on disasters rather than other fatal incidents—far better data are available.

Since 1960, in the U.S., the U.K., and Australia, whenever there has been a coal mine disaster, a thorough investigation and detailed public report by the government on its causes has ensued. The only exception was the Wyee disaster, in 1966, when five men lost their lives in a roof fall. The Wyee mine was owned by the Government of New South Wales. Apparently the New South Wales Cabinet decided against an open, public enquiry and report on the causes of the death, when it became clear that such an enquiry could lead to demands for the government to launch prosecutions against one of its own entities.

The process that leads to a government report on a disaster is variable. In New South Wales, a judge is appointed to conduct public hearings and make findings in the style of a Royal Commission. In Queensland, the Mining Warden (a magistrate) sits with a panel of mining experts in public hearings. Independent tribunals can be appointed by the Parliament, in Britain, to conduct public hearings and report on disasters (as in the Aberfan disaster). The more common recourse, however, is a detailed report by the Chief Inspector of Mines and Quarries. Sometimes public hearings on the disaster are conducted, sometimes evidence is gathered in camera. In any case, in addition (as in Australia), an inquest is always conducted in public, and the Chief Inspector will often draw upon evidence from the inquest in writing his report. In the United States, the Mine Safety and Health Administrations (formerly the Bureau of Mines) conducts investigations, which do not always draw upon public hearings. Often the report is written after hearings conducted by a congressional subcommittee or called by the relevant state legislature, but often they are not. The Belgian and Rhodesian reports included in this study were written by independent committees of enquiry that held public hearings.

We need not describe in greater detail the way in which investigations of disasters are conducted in the different countries. Suffice it to say that in all these countries public concern about the disasters has been great enough to produce vigorous investigation and public reporting. The same could not be said about single mine fatalities or nonfatal accidents.

## THE SAMPLE

The sample is in fact very nearly the whole population of disasters in the United States, Great Britain, and Australia between 1960 and 1981. In Britain, reports were available for all disasters during this period. Only the Wyee disaster was missing for Australia, and in the U.S., only four were unavailable, because of court orders or failure to complete the reports. Thirty-four disasters since 1960 were therefore available from these three countries. I was also able to obtain reports on the only major disasters in Zimbabwe and Belgium during recent times—the Wankie disaster, in which 425 lost their lives, and the Bois-de-Cazier disaster, which killed 262—and these were added to the study both to internationalize the data further and because of their enormous importance. Finally, three incidents were added that luckily do not fit our definition of disasters because only four persons perished in each. These were the Bulli Mine fire in Australia, the P and P coal Company, No. 2 Mine explosion in the U.S., and the Cardowan explosion in Britain. There may well have been other incidents in these countries during which four persons have perished, but these three were major incidents in which the loss of life could easily have been much greater, and for which the reports were available. The thirty-nine disasters are listed in table 1. It shows that most of them resulted from explosions (either of methane or of coal dust, or both, or in one case, simply an explosion of explosives).

TABLE 1

THE THIRTY-NINE DISASTERS*

| Year | Mine | Place | No. of Deaths | Cause |
|------|------|-------|---------------|-------|
| *United States* | | | | |
| 1960 | Pine Creek No. 22 | West Virginia | 18 | Mine Fire |
| 1961 | Viking | Indiana | 22 | Explosion |
| 1962 | Blue Blaze No. 2 | Illinois | 11 | Explosion |
| 1962 | Robena No. 3 | Pennsylvania | 37 | Explosion |
| 1963 | Compass No. 2 | West Virginia | 22 | Explosion |
| 1963 | Carbon Fuel No. 2 | Utah | 9 | Explosion |
| 1965 | Kline No. 2a | Tennessee | 5 | Explosion |
| 1965 | Mars No. 2 | West Virginia | 7 | Fire/ Explosion |
| 1965 | Dutch Creek No. 2 | Colorado | 9 | Explosion |
| 1966 | Dora No. 2 | Pennsylvania | 5 | Suffocation |
| 1966 | Siltix | West Virginia | 7 | Explosion |
| 1968 | River Queen No. 1 | Kentucky | 9 | Explosion |
| 1970 | Finley Nos. 15-16 | Kentucky | 38 | Explosion |
| 1972 | Buffalo Creek | West Virginia | 125 | Tip Slide |
| 1972 | Blacksville No. 1 | West Virginia | 9 | Fire |
| 1972 | Itmann No. 3 | West Virginia | 5 | Explosion |
| 1977 | P and P No. 2 | Virginia | 4 | Explosion |
| 1978 | Moss No. 3 | Virginia | 5 | Inundation (blackdamp) |
| 1981 | Dutch Creek No. 1 | Colorado | 15 | Explosion |
| *United Kingdom* | | | | |
| 1960 | Cardowan | Lanarkshire | 4 | Explosion |
| 1960 | Six Bells | Monmouthshire | 45 | Explosion |
| 1962 | Hampton Valley | Lancashire | 19 | Explosion |
| 1962 | Tower | Glamorganshire | 9 | Explosion |
| 1965 | Cambrian | Glamorgan | 31 | Explosion |
| 1966 | Merthyr Vale | Aberfan | 144 | Tip Slide |
| 1967 | Michael | Fife | 9 | Fire |

| 1971 | Cynheidre/ Pentremawr | Carmarthenshire | 6 | Outburst (Coal & Firedamp) |
| 1973 | Lofthouse | Yorkshire | 7 | Inundation (water) |
| 1973 | Seafield | Fife | 5 | Roof Fall |
| 1973 | Markham | Derbyshire | 18 | Overwind & Cage Crash |
| 1975 | Houghton Main | South Yorkshire | 5 | Explosion |
| 1978 | Bentley | South Yorkshire | 7 | Train Crash |
| 1979 | Golborne | Manchester | 10 | Explosion |

*Australia*

| 1965 | Bulli | New South Wales | 4 | Fire |
| 1972 | Box Flat | Queensland | 17 | Explosion |
| 1975 | Kianga No. 1 | Queensland | 13 | Explosion |
| 1979 | Appin | New South Wales | 14 | Explosion |

*Belguim*

| 1956 | Bois-de-Cazier | Marcinelle | 262 | Fire |

*Zimbabwe*

| 1973 | Wankie | Middle Zambesi Valley | 425 | Explosion |

*NOTE: See Appendix for the citations of the disaster reports.

METHODOLOGY

It was not the goal of this study to find out what technical deficiencies led to disasters (e.g., whether defective safety lamps or frayed cables more oftten ignited gas), but what managerial deficiencies were the culprits. For obvious reasons, given the purpose of this book, there was nevertheless an interest in technical matters to the extent necessary for establishing whether a violation of law caused the disaster. The first task, then, was to discover from each of the reports whether the investigation found offenses and whether these offenses contributed to the disaster, or if the disaster was an uncontrollable "act of nature," which no level of compliance with law could have prevented.

In addressing these questions, the finding in the official government report is our raw data. If the report does not mention any violations, then no violations are counted. This is so, even though a report may have been defective in failing to point to violations. In some specific instances, I believe, from a reading of transcripts of evidence and my own interviews, that government reports are in error. However, there are real problems when a nonauthoritative source (this author) begins to modify selectively a collection of authoritative reports. Readers will ask unanswerable questions about why some findings were changed in one direction while others were not changed in the opposite direction. Although the government reports are wrong some of the time, they will be right more often than would any individual researcher with limited resources and technical capacity and no ability to visit the sites of the disasters. Second-guessing government findings would introduce error as well as correcting it and would certainly give the findings less credibility.

The goal of this research was an ambitious one—and one that was not completely fulfilled—to ascertain what organizational defects and which individuals within the organization were responsible for each disaster. If we knew this, we could better judge whether punishment was the best way of preventing disasters, and if so, which individuals or organizational units should receive the punishment.

After reading about half the reports for the first time, it became clear that a small number of organizational defects were repeatedly showing up as responsible for the disasters. A couple of less common problems emerged later, at which point it was necessary to reread all the reports and mark on the score sheet the cases where these problems were found to be a cause of the disaster. The organizational defects found to be most frequently cited (after the author personally read all thirty-nine reports several times) were: the lack of a plan to deal with a hazard, or poor planning; a generalized pattern of inattention, or sloppiness, in relation to safety matters; poor communication, or reporting systems; inadequate training; and inadequate definition of responsibilities. Before going on to the detailed findings on these organizational defects, however, we will first discuss the findings on whether violations of law were responsible for the disasters.

## Violations as Causes

The first question to be addressed was whether the reports found that the disasters could not have been prevented even if the law had been observed. That is, in each case did the report find that the mine had a basically good compliance system, and that the disaster was instead caused by either individual human error or forces of nature beyond human prediction and control? In no case was there such an explicit finding, but in two cases this was the clear implication of the reports. The 1972 Itmann No. 3 explosion was found to have occurred despite an adequate ventilation system; the investigation uncovered no violations of the Mine Safety and Health Act in the section of the mine where the explosion occurred. Rather, the finding was that there had been an unexpected and sudden liberation of methane, possibly from extraordinary pressure from adjacent strata, and this was ignited by an electric arc generated by a bus. A similar "unstoppable forces of nature" finding was made concerning the 1971 Cynheidre/Pentremawr outburst, in West Wales. An "outburst" is a sudden gushing of soft coal and methane from geological pressure at the coal face; miners are buried in the outburst of coal. In this case, the report clearly implied that no preventative measure would have saved either the six killed or the sixty-nine others who suffered varying degrees of asphyxia.

Clearly, two out of thirty-nine reports with this kind of finding is not many. By contrast, in thirty-three of the thirty-nine reports, serious violations of safety laws were reported as having been uncovered in the investigations. "Serious" in this context means violations *commented on* in the report as *serious*, rather than unimportant technicalities. In six of the fourteen British investigations, there was no report of a violation of the law, though in some of these, violations of approved codes of practice were found. This difference between violations of the law and of approved practice is not likely to be explained by the fact that British disasters are less likely to be caused by illegality than by the tendency of the writing style in British reports to sidestep legal questions and concentrate on technical detail in a way that does not relate to the content of the law. Even when the chief inspector decides that a determination is required as to whether an offense did cause a given accident, he still often manages to sidestep the issue, as illustrated by the following two paragraphs from the report on the overwinding accident at Markham Colliery, in which eighteen men perished when the cage in which they were traveling plunged to the bottom of the pit:

> 61. Section 81 of the Mines and Quarries Act 1954, requires that "All parts and working gear, whether fixed or movable, including the anchoring and fixing appliances, of all machinery and apparatus used as, or forming, part of the equipment of a mine, and all foundations in or to which any such appliances are anchored or fixed shall be of good construction, suitable material, adequate strength and free from patent defect, and shall be properly maintained." It should be considered whether, in the circumstances, there has been a contravention of this section which it would have been practicable to avoid or prevent.
> 62. Evidence given at the Inquiry showed that the trunnion bearing did not operate as designed and caused fluctuating stresses to be induced in the centre rod of the spring nest which it could not sustain. R. Jeffrey, Senior Scientific Officer, Safety in Mines Research Establishment, said that any crack which had penetrated to a depth of ⅜ inch could have been detected by ultrasonic tests without removing the rod, but not by visual examination with the unaided eye. There is uncertainty as to the frequency of ultrasonic testing which would have been necessary to discover the main pre-existing crack as its rate of propagation beyond ⅜ inch could not be established. It appears that the persons at the colliery having responsibility for maintenance had no reason to suspect that the rod was overstressed and, therefore, did not appreciate the need for special methods of testing. Nevertheless, the rod was shown to be of inadequate strength for the stresses induced and the main pre-existing crack could have been found by available means of testing. Also, there was a precedent for this type of failure in a similar rod at Ollerton Colliery in 1961.

With this kind of ambiguity, the report could only be scored as having no finding of a violation.

The second question was, How many of the reports contained a finding that the violations uncovered by the investigations were a contributing factor in the deaths? A violation as a "contributing factor" means a finding that at least some of the deaths would probably not have occurred had the laws been complied with. In most disasters there are many contributing factors, only some of which are violations of law. But even when a violaiton was only one of many causes of a disaster, compliance with the law would often have prevented the disaster. For example, consider a common situation, an explosion in which one of the causes was a sudden increase in the liberation of methane from the mine face, another was a mechanical breakdown in a fan, and a third was that, in testing the fan, it was turned on illegally before its flameproof casing was secured. An electric spark from the fan then ignited the explosion. Notwithstanding the first two fundamental causes, which were not violations of law, if the law had not been violated by illegally turning on the fan, the explosion would not have occurred.

In some of the disasters studied, several violations were causes, and compliance with respect to any one of these violations might have prevented the disaster. Consider, for example, the 1978 train crash at Bentley Colliery, South Yorkshire, in which seven men died and three were seriously injured. The train ran out of control when moving down a steep slope in the mine. First, this slope was steeper than permitted by the regulations. Second, the driver drove down the slope in second gear, contrary to the rules and in disregard of notices requiring first gear. Third, the disaster might have been prevented had emergency brakes been applied on the carriages, but the guard was not in the correct position from which to apply the brakes. This problem, in turn, might have been avoided had the man concerned been trained or authorized to act in this job. In spite of all these offenses, the accident would have been prevented if an arrestor positioned for the very purpose of stopping runaway trains had not been turned off. The arrestor consisted of an impact head that engaged twelve successive pairs of friction clamps. It had been turned off to save the trouble of pulling a lever to lower its impact head below the rail level to allow the train to pass over unhindered.

Then there were violations that served to exacerbate a disaster. The most common was a failure to meet legal requirements for covering tunnels with rock dust to prevent a spreading of explosions through ignition of coal dust on the floor and sides of the tunnel; compliance would have produced a less extensive explosion. Another example was a failure properly to maintain the rescue equipment required by law at the mine, resulting in the rescue of fewer miners than would otherwise have occurred.

For twenty-five of the thirty-nine disasters (64 percent), it was found that a violation or violations were contributing factors (either as causes without which the disaster would not have occurred or as factors without which the disaster probably would have been less serious). Moreover, some of the fourteen cases for which no violation contributing to the disaster was found must be treated with some scepticism. The Wankie disaster, in Zimbabwe (then Rhodesia), is the best example. Such a massive loss of life (425 dead) would not have occurred had the mine been rock-dusted. But rock dust was not used at all at the mines; incredibly, Rhodesian law did not require its use. Where there is no law, there can be no violation. There were also at least three cases that had contributing factors not judged as offenses at the time of the disasters but that would be violations under the present laws of the country involved.

Then, too, for several disasters there was some ambiguity about whether a contributing factor was a violation. An example is Yorkshire's Lofthourse Colliery inundation in which seven men were killed when the mind was flooded with water after unexpectedly cutting into old workings. The manager was not aware that the miners might cut into flooded old workings because of an inadequate job of checking old mine maps for the district. Yet the manager, under Section 75 of the Mines and Quarries Act 1954, had a duty:

> (a) of taking such steps as may be necessary for securing that he is at all material times in possession of all information which indicates or tends to indicate the presence or absence, in the vicinity of any workings carried on or proposed to be carried on in the mine, of: (i) any disused workings (whether mine workings or not); (ii) any rock or stratum containing or likely to contain water (whether dispersed or

in natural cavities); (iii) any peat, moss, sand, gravel, silt or other material that is likely to flow when wet; and

(b) of taking such steps as may be necessary for the purpose of substantiating any such information which comes into his possession (whether in consequence of the discharge of the duty imposed upon him by the foregoing paragraph or not).

While the chief inspector did not find that the manager had failed to perform this duty, he did find that the summary report from the surveyors on old workings in the area was "a bald and restricted statement in which no attempt was made to present the documents investigated" (p. 15), that the manager did not examine this information, and that he signed the plan following inadequate oral discussions which did not review the thinking behind the summary report and were not recorded. Clearly, one could interpret the report to mean that the manager and others with statutory duties failed to meet them, and that this caused the disaster.

Also, among the fourteen reports for which no violation was found to have contributed to the disaster, there are four for which violations were found to have been *possible* contributing factors. Again, Wankie is a good example:

In our opinion, contraventions of the Mining Regulations, 1961, and of the Explosives (Licensing and Use) Regulations, 1970, occurred before the disaster. ...Any of them might have caused the explosion. However, as the field of possible causes is considerable, none might in fact have contributed to the disaster (p. 39).

Some of the other cases were more specific. For example, in the Robena No. 3 explosion of 1962, in Pennsylvania, four possible sources of ignition were identified—only one of which involved a violation of the law. Another common type of finding was in the report into the Box Flat disaster, in Queensland:

Turning to the question of stone dusting, it is observed that the requirements of the Acts and Regulations were not complied with. Whether full compliance would have made any difference in the circumstances is not known... (p. 82).

WHO WAS RESPONSIBLE?

In those twenty-five disasters in which an offense was a contributing factor, what were the findings about the responsibility of individuals or collectivities? In only four was the corporation as an organization specifically named as responsible. However, it was generally implicit in the U.S. reports that the companies were responsible, and, indeed, it was the companies that were invariably fined for the violations recorded in the reports.

The Haughton Main explosion enquiry of 1975, in South Yorkshire, was unique in that blame was placed on the Safety Department as a collectivity and the Mechanical Engineering Department as a collectivity, as well as on a number of individuals.

Reports from all countries tended to avoid determinations of individual blame, particularly in the United States. In six reports, persons at the foreman/deputy level were listed as responsible for violations. This was clearly the most common level of the organization at which individuals were held blameworthy. In no cases were normal miners singled out for blame (see table 2).

Table 2

Number of Disaster Reports Holding Individuals Responsible for a Contributing Violation

| Organizational Level | Number of Reports |
| --- | --- |
| Foreman/deputy | 6 |
| Mine manager | 3 |
| Engineer | 2 |
| Manager safety department | 2 |
| Electrician | 2 |
| Undermanager | 1 |
| Surveyor | 1 |
| Locomotive driver/guard | 1 |

The data on individual responsibility is certainly thin. However, it at least raises the question of whether systems of individual accountability for observing the law should focus particular attention on the foreman/deputy.

Not surprisingly, given the normal control of regulatory agencies over the investigative process, in only three reports was there a clear finding that inspectors had both known about the hazardous practices uncovered in the investigation and permitted them to continue. In two other cases there were adverse findings that inspectors had failed to notice problems they should have noticed.

## Uncertainty about the Causes

A major question when considering the effectiveness of a punitive approach to regulation is whether certainty as to a disaster's causes is possible. Clearly, it is difficult to punish people and organizations for what happened if one cannot be sure of what transpired. In fifteen of the thirty-nine reports there was an expression of some significant uncertainty as to the causes of the disaster.

After explosions, it is particularly difficult to reconstruct causes because the blast itself destroys so much evidence. Typical are the findings on the Finley explosion, in Hyden, Kentucky, in 1970:

> It is the conclusion of the Bureau of Mines that the explosion occurred when coal dust was thrown into suspension and ignited by Primacord or by permissible explosives used in a nonpermissible manner or by use of nonpermissible explosives during the blasting of roof rock for a loading point (boom hole). Excessive accumulations of coal dust, and inadequate applications of rock dust in parts of Nos. 15 and 16 mines permitted propagation of the explosion throughout the mines (p. 25).

This kind of uncertainty in 38 percent of the disasters means that it would be impossible in these cases to impose retributive sentences in keeping with the enormity of the disasters. However, it is still possible to punish those responsible for offenses on the grounds that they had illegally taken unnecessary risks, irrespective of whether those risks or others caused the disaster.

## LACK OF PLANNING AS A CAUSE

In twenty-two reports the lack of a plan to deal with hazards, or poor planning, was found to have contributed to the disaster. Unfavorable attention was drawn to poor planning in some other cases, but here the planning inadequacy was not found to have contributed to the disaster. For example, in the Michael Colliery disaster of 1967, the chief inspector criticized the lack of an emergency and rescue plan for the delayed rescue efforts, but those who lost their lives would have perished even without the delay.

Mining can be a complex activity. As a consequence, problems can arise when people decide to start a job and work out the details of their procedure as they go along. The report on the explosion at Cardowan Colliery, near Glasgow, is typical of the twenty-two cases in which poor planning was identified as a fundamental cause of the catastrophe. The explosion occurred during operations following the pulling down of stoppings, erected some months previously to seal off an underground fire.

The exploration and, in particular, the re-establishment of the ventilation appears to have been carried out on an *ad hoc* basis and there seems to have been little sense of urgency. The plans for breaching the stoppings in the drifts were excellent and had been laid down beforehand in detail, but what was to happen after that seems to have been left to be decided upon in the light of the state of affairs at the time. The result was delay, amounting to days, in restoring the ventilation and in clearing the accessible workings of firedamp. The contours of the area were known and flooding of the airways in the "swilley" should have been anticipated and the necessary pumping gear should have been ready. It does not seem to have been realised that once fresh air reached the top of the drifts it would flow downhill under the lighter, less dense, firedamp and that it would not all enter the return where the doors at No. 1 East Air Crossing had been left open in January. For lack of a definite plan to restore the ventilation right into No. 4 East on a fixed time-table of *hours*, together with the provision of the necessary materials to seal off leakage at the various connections between intake and return, a period of no less than nine days elapsed before anything approaching a sufficient volume of air was supplied to begin to clear the gas from the rise side of No. 4 East District. Had the plan been to clear the roadways in stages (as it might very well have been in view of the very large volume of firedamp involved) this could have been settled beforehand; the points at which temporary stoppings were to be erected decided upon and the necessary materials provided (pp. 14-15).

The recommendations of the chief inspector in this report were mirrored in many of the other reports:

All phases of work involved in re-opening a district which has been sealed off following a fire, and in restoring conditions to normal, should be covered by a detailed plan, drawn up in consultation with all interested parties. This plan should not be varied materially, except as may be necessary to deal with an emergency, without fresh consultation with those parties. The scheme should provide for the appointment fo one senior official to be in charge of all operations on each shift, and for the provision of ample supplies of necessary materials... (p. 17).

A more recent American example is provided by the inundation of blackdamp (oxygen-deficient air) that occurred when a continuous miner digging a drainway cut into a mined-out, inaccessible abandoned area of the same Moss No. 3 mine, in Virginia, in 1978. The Mine Safety and Health Administration (MSHA) investigation concluded:

...the planning of the Drainway project by mine management and the Safety Department was inadequate and incomplete because: (a) due consideration was not given to the possibility of the abandoned 1 Right area containing blackdamp; (b) the final plan contained no provisions or safety precautions that would permit the Drainway to penetrate the abandoned area in a safe manner and under controlled conditions in the event blackdamp was encountered; (c) mine management did not discuss the possibility of the abandoned 1 Right area containing blackdamp with the employees at the Drainway prior to or during development of the Drainway entry and four workmen were caught unaware by the inrush of blackdamp when the entry cut through into the abandoned area; and (d) during the investigation the Drainway ventilation system was found to be inadequate which indicates that the ventilation system was a product of inadequate and incomplete planning (pp. 22-23).

Quite a number of the disasters occurred in mines that had soundly thought-out plans for dealing with normal mining routines, but reacted to unusual circumstances by muddling through instead of by redesigning a special plan of attack. The finding of the Robena No. 3 explosion report of 1962 is typical. The disaster occurred because "the mining sequence was changed temporarily from an established and simple routine to a more complex system that made ventilation more difficult to direct and control" (p. 24).

The significance of poor planning as a cause of mine disasters has implications for determining the kinds of conduct that should be punished in mine safety enforcement. Regulatory strategies are already in use in various parts of the world that take punitive action for failure to maintain and abide by plans written at the mine, rather than for failure to comply with rules written by the regulatory agency.

## POOR COMMUNICATIONS AS A CAUSE

Seventeen of the reports found that inadequate communication of hazards, or weak reporting systems, were fundamental causes of the disaster. There was considerable overlap with the factor discussed in the last section—many disasters were caused by defective plans defectively communicated. In other cases, like the 1979 Appin explosion, in New South Wales, the plan for changing the direction of airflow in the ventilation system was adequate enough, but communication, down the management hierarchy, of what had to be done was defective. The confusion cost fourteen men their lives.

The 1966 Aberfan disaster, in Wales, is the classic illustration of inadequate communication systems as a fundamental cause. At least 144 persons were killed, 116 of them schoolchildren, when a huge tip of mine waste that had been dumped over a watercourse was washed down a valley. There was no excuse for this disaster; responsible officials should have been aware that slipping of tips had already occurred in the same area and even at the same mine. With adequate communication of the dangers highlighted by these earlier incidents, responsible officials would have acted to prevent the disaster. The tribunal concluded:

> ...a complete overhaul in the system of intercommunication (both vertically and horizontally) between the various departments or levels of the [National Coal] Board is urgently needed.... One of the strangest features which emerged from the evidence was the frequent absence of adequate lines of communication and of failure in collaboration between Board colleagues with linked responsibilities (p. 123).

The report then went on to catalogue an enormous number of critical communication breakdowns. I have edited many of them out of the following quotation. Even so, it is quite lengthy enough to give the reader the flavor of communication problems.

> Mr. W. V. Sheppard, Director-General of Production, although head of the department at Board headquarters which was responsible for all tips, told the Tribunal that not only was he wholly without knowledge of the 1939 and 1944 slides (which preceded nationalization) but also that he had never heard of the 1963 slip or of the 1965 incident at Tymawr, or of any difficulty with tailings, and that he had had his attention drawn to no literature or memoranda dealing with tip slides or similar incidents. No record of any untoward tip happenings in the South Western Division seems ever to have found its way to London.
>
> The elaborate site investigation conducted at Nantgarw Colliery in 1958, and the machinery set up by Mr. Blackmore, manager of No. 3 Area in 1965, to investigate the Tymawr incident seemed to have been largely, if not entirely, unknown in other Areas....
>
> In August, 1963, the Merthyr Tydfil Borough Council wrote expressing to Mr. D. L. Roberts (Area Mechanical Engineer) their fear of the serious position that would arise if the tip on Merthyr Mountain moved. At no time did Mr. Roberts disclose this correspondence to the Colliery Manager, who testified, rightly or wrongly, that, had he known of it, "I would go and get a complete investigation as to what the letter was alluding to...."

The Area General Manager, Mr. Wright, was likewise left uninformed about this correspondence and, having seen and heard him, we think it probable that he would have ordered an immediate investigation had he had knowledge of it....

Although Mr. D. L. J. Powell sent his 1965 memorandum from Division to Area Mechanical and Civil Engineers, he sent no copy to the Area General Manager (Mr. Wright), who had no knowledge of its existence, knew nothing of the tips inspection (such as it was) conducted in purported compliance there-with.... Had the Area General Manager been kept informed, he might well have realised (at a time when the alarming Tymawr incident remained fresh) that the Area Civil Engineer had made no inspection at all.

The breakdown in communication between the Area Mechanical and Civil Engineers...in relation to the inspection of Tip 7...had dire consequences. As Mr. Geoffrey Morgan accepted, had there been such collaboration, the probability is that there would have been no Aberfan disaster (pp. 123-4).

An implication of the widespread involvement of defective communication and reporting in disasters may be that regulatory agencies should direct considerable enforcement effort at the responsibility to know about and to report dangers.

A PATTERN OF INATTENTION AS A CAUSE

The third recurrent factor was a finding that the deficiencies leading to a disaster were part of an ongoing pattern of inattention and sloppiness in safety matters. This seems a vague notion, but in all its vagueness it was pointed to in seventeen reports. Moreover, we will see later that it is a vague notion that can have quite concrete policy implications.

I suspect this empirical finding understates the extent to which disasters are the ultimate result of an ongoing pattern of disregard for safety. I say this because in so many of the reports with no finding of a pattern, there was nevertheless such a catalogue of safety violations and unsafe practices that one wondered why it was not concluded that these companies had a sloppy attitude toward safety.

Chapter 5 includes a discussion of the U.S. Mine Safety and Health Act provisions for determining that a mine has a pattern of violations. This provision was enacted when Congress discovered that mines at which there had been disasters frequently had an extraordinary history of safety violations, often of the same type that caused the disaster. The findings discussed here reinforce the empirical justification for such a legal provision.

## Inadequate Training as a Cause

In fourteen reports, inadequate training of one sort or another was given as an important contributor to the disaster. This finding could be taken to imply that mandatory training requirements ought to be enforced. Equally, as discussed in chapter 1, it could imply that resources ought to be diverted away from law enforcement and toward training. Since this important question is beyond the scope of this book, we will do no more than draw attention to this finding in the hope that others will take up the issue of the relevance of inadequate training to disasters.

## Unclear Lines of Responsibility as a Cause

The final recurring organizational defect in the disaster reports was clearly the least common. Confusion over who was responsible for what was reported as a fundamental cause in only five reports. However, other reports contained tangential references to failures in deliniating precisely who was accountable for which aspects of safety performance. Moreover, there were additional cases in which unclear responsibility was a problem, but, fortuitously, it did not contribute to the loss of life. For example, for ninety minutes after the fire broke out at Fife Colliery, in 1967, the rescue brigade was not summoned because no one felt he had the authority to declare a state of emergency:

> Clearly there was uncertainty among those officials immediately concerned...about the exercise of authority to declare a formal state of emergency at the colliery. Lister, upon whom fell the brunt of having to deal with the emergency when it was first reported to him, certainly did not regard himself as authorised to

initiate the full emergency procedure. Webster, who was at a remote point underground and who was not really in a position to judge how serious matters were, referred Lister to Simpson, an under-manager, or Nicol, the deputy manager. Even Nicol hesitated to take the positive step of declaring an emergency and it was not until 5:30 a.m., some two hours after the finding of the fire, that a general emergency was in fact declared on the instructions of J. S. Wilson, production manager.... [T]he presentation of the organization chart lacked the clarity which I regard as absolutely essential (p. 26).

Luckily in this instance, the delay in calling the rescue brigade, setting up an incident room, and other emergency measures had no effect on the ultimate loss of life.

Problems of diffused and confused accountability are not straightforward matters to illustrate. Therefore the following extract indicating how inadequate lines of responsibility contributed to the Aberfan tip slide, is lengthy. It shows how, contrary to instructions, Mr. Exley, who as a civil engineer was the person most likely to grasp the gravity of the risk, was excluded from collaborating on an inspection of the tip site.

A contributory factor may have been that when Mr. Exley was first appointed in 1958 he did not receive, as he should have done, notice defining his duties as required by Section 1 of the Mines and Quarries Act, 1954. He received his notice in this behalf three years later, in May, 1961, and it informed him that he was responsible for, "technical engineering control of all surface buildings and structures at mines in the No. 4 Area...." When Mr. Roberts, however joined the Area as Mechanical Engineer in 1960 he received his notice in August of that year. It informed him that he was responsible for "technical engineering control of all mechanical apparatus *and of all surface buildings and structures* at mines in the No. 4 Area...." Although Mr. Exley had been longer in post in the Area, Mr. Roberts was the first to receive his notice under Section 1 and he was given no amending notice when Mr. Exley received his notice. Each, therefore, had reason to believe that he was responsible for "civil engineering matters...."

Mr. Roberts, who said that at times he and Mr. Exley were "at loggerheads," asserted that he assumed that he and Mr. Exley would be sending in independent reports. Mr. Exley, on the other hand, denied any estrangement and claimed that Mr. Roberts expressly told him that he would himself see to the report and so Mr. Exley left it at that. We do not believe either of these versions. We think that the truth is that, it being traditional that tips were the responsibility of mechanical engineers at the various levels, Mr. Roberts deliberately ignored the fact that Mr. Exley was one of the addressees of the Powell letter.... Mr. Exley, on the other hand, heavily overshadowed by the far more dominant Mr. Roberts, never consulted him about the matter at all, and took it for granted that his assistance was not welcome to Mr. Roberts, whatever Mr. Powell intended.

Be that as it may, there is unanimity of views among the higher officials of the National Coal Board...that collaboration between mechanical and civil engineers was essential to the checking of tip stability.... The absence of their collaboration was literally disastrous in its consequences, for...had they co-operated the disaster of 1966 would probably never have happened (p. 76).

## CARBON-COPY DISASTERS

In reading the thirty-nine disaster reports, I often became confused that perhaps I was reading one I had already read, so many of the disasters in different parts of the world were so similar. The first one I read was that of the 1979 Appin disaster, in New South Wales. There, Judge Goran found that the explosion, which killed fourteen men, had been ignited when an electrician tested a fan by turning it on before the flameproof cover on the fan was secured by tightening ten bolts. A spark from the fan blew up the mine. I later read the 1965 report on the explosion of the Cambrian Colliery, in Glamorgan. It was the same depressing story—thirty-one had died when an electrician tested a gate-end switch without going through the tedium of tightening the ten bolts to render the switch flameproof.

An explosion requires both an ignition source and an accumulation of explosive gas. Judge Goran was scathing in his attack on the owners of the Appin mine, Australian Iron and Steel (a subsidiary of BHP), for their appraoch to methane in the mine:

What in fact was allowed to happen was the growth of a philosophical attitude towards methane as a fact of life. It was a nuisance, it could hold up production in working places, but it was not a matter of great concern in standing places where the possibility of ignition was remote. The officials had their own view of when methane gas was permissible. It differed from the standard of the Act (p. 86).

What a shock it was, then, to read the Bulli disaster report, written in 1965 by the same judge about the same company, concerning a mine that was only a few miles from Appin:

> This basic hiatus in the reasoning of the management led it to treat noxious gas as a gas which could be tolerated providing it was not forming at or near the fact itself. In the shunt it tolerated concentrations of noxious gas which drew complaints from the workmen. Its method of dealing with this gas was a mere improvisation for which no justification could be found in mining practice and which was dangerous in the extreme.... This can only be called a cavalier attitude on the part of management and the Deputies towards the problem of gas at Bulli Colliery (p. 37).

Judge Goran was prophetic in concluding:

> ...I must issue a warning that the tragedy which occurred on the morning of the 9th November 1965 at Bulli Colliery can be repeated at this and other collieries with possibly far more disastrous results unless the lessons learned at this Inquiry become rooted in the minds of all men who work in mines, both management and employees (p. 37).

Bulli, incidentally, had been the scene of Australia's second worst coal-mining disaster, in 1887, when eighty-one perished in an explosion.

The same problem is evident in the United States. Seven men died, in 1965, at the Mars No. 2 Mine of the Clinchfield Coal Company, in a fire and explosion started when a continuous mining machine being trammed along a tunnel contacted energized trolley wires. The problem was that this tunnel had insufficient clearance for moving bulky equipment without a risk of contact with the overhead wires. Employees testifying at the official hearing said that equipment being trammed had come in contact with the power circuit in this tunnel on three different occasions prior to the disaster. The required preventative measure, according to the report, was simple:

When large bulky equipment is being moved along haulageways having energized power wires and scant clearances, the power wires shall be deenergized in the area through which the equipment is being moved (p. 17).

The men died in this mine because for ventilation they depended on air passing through the tunnel in which the fire was started by the arcing of the contacted trolley wires. Thus, the second important recommendation of the Bureau of Mines report was:

Men should not be permitted to work in areas that are ventilated with air coursed over large bulky equipment being moved along travel routes having scant clearances and energized power wires (p. 17).

Lo and behold, less than seven years later in the same state, nine miners were entombed after they were caught working inby a fire started by a continuous miner, being transported along the tunnel that supplied them with air, when it made contact with energized trolley wires. After the Mars No. 2 disaster it was made an offense under West Virginia mining law to move equipment with persons inby the equipment in the same ventilating current; after this second disaster at the Consolidation Coal Company's Blacksville No. 1 Mine, a similar offense was created under the Federal Mine Safety and Health Act. But the Blacksville No. 1 disaster was a carbon copy of the Mars No. 2 fire and explosion in more ways than one. The Bureau of Mines Interim Report on Blacksville repeats the following tale of negligence:

Movement of this miner was begun on the day shift and continued on the afternoon shift despite knowledge of contact with the trolley wire and entry roof by a loading machine while it was being moved on the morning shift in that same entry on the same day—contact also incurred because of the tight clearance (p. 2).

The most lamentable examples of failure to learn the lessons of the past, however, are in the area of slides of tip heaps of mine waste. The Aberfan disaster shocked the world because

116 of the victims were schoolchildren, mostly between 7 and 10 years old. Yet even more shocking was the way the tribunal report showed that slides of mine tips had been common in South Wales, and that three had occurred at Aberfan itself within a span of 25 years, and a fourth less than 5 miles away, 27 years earlier. In this latter incident, 180,000 tons of sludge slid down a valley, taking 500 feet of railway line with it and causing considerable further property damage. Fortunately, none of these previous slides in South Wales had ever claimed a life. But managers of the National Coal Board should have realized that this had been a matter of extreme good fortune.

Six years later came the Buffalo Creek disaster, in West Virginia, when at least 125 people lost their lives as another glacier of coal mine sludge descended on a helpless valley. The Aberfan disaster had triggered investigations, sponsored by the U.S. Geological Survey, into similar refuse piles in America. The Buffalo Creek refuse pile, which served also as a dam, was subjected to unfavorable comment in this survey. But between the 1966 survey and 1972, far from being rectified, the situation was made far worse by the dumping of more waste and the buildup of a bigger and bigger body of water behind three makeshift dams. The Pittston Company, which owned the Buffalo Creek mine, should also have learned the lessons of Aberfan, because top executives of the company were sent a copy of the National Coal Board's booklet, *Spoil Heaps and Lagoons*. This had been prepared to ensure against another Aberfan, and it clearly points out that Pittston's situation at Buffalo Creek was extremely dangerous. Moreover, Mr. Spotte, president of the Pittston Coal Group, had corresponded with a Mr. Round, a National Coal Board expert on tip slides, because, in Mr. Round's words, "Irvin Spotte was operating a number of situations which had the potential for similar disaster [to Aberfan]" (Stern 1976, p. 202).

All this must have been highlighted further for Pittston when, five months after Aberfan, another Pittston refuse pile dam in West Virginia gave way under pressure of heavy rainwater, damaging a substantial land area and filling one house with coal refuse. At about the same time, one of the refuse pile dams at Buffalo Creek collapsed into another larger dam below it. Back in 1955, another of Pittston's refuse tip dams had collapsed in Virginia, sweeping away two houses. Finally, Pittston should have been aware of a number of previous tip slides that had occurred in West Virginia since the 1920s, including one that took seven lives. In spite of all of these events that should have made Pittston aware of the dangers, the company did not act to make the refuse pile safe as recommended in the National Coal Board document, *Spoil Heaps and Lagoons*. Even the following prophetic letter to the governor of West Virginia from a subsequent victim of the Buffalo Creek disaster, Mrs. Pearl Woodrum, did not stir action:

Dear Sir,
I live three miles above Lorado. I'm writing you about a big dam of water above us. The coal co. has dumped a big pile of slate about 4 or 5 hundred feet high. The water behind it is about 400 feet deep and it is like a river. It is endangering our homes & lives. There are over 20 families here & they own their homes. Please send someone here to see the water & see how dangerous it is. Every time it rains it scares everyone to death. We are all afraid we will be washed away and drowned. They just keep dumping slate and slush in the water and making it more dangerous every day. Please let me hear from you at once and please for God's sake have the dump and water destroyed. Our lives are in danger.

As an aside, it is worth mentioning that four of the nineteen U.S. disasters in this study were at Pittston mines—appalling for a company that mines only about 2 percent of the country's coal.

Before completing this chapter, I became aware of a report in the *Multinational Monitor* (November 1982, p. 9) of the estimated loss of more than 200 lives in Liberia when "a rain-swollen tailings dam burst and swamped" the mining camp below the dam. Investigation of the disaster is still proceeding at the time of writing.

Some of these carbon-copy disasters could have been avoided with better communication of the lessons of previous disasters. But the Buffalo Creek disaster illustrates vividly that better communication concerning hazards is not enough—Pittston knew the hazards and chose to ignore them. Pittston benefited from all the gentle coaxing and communication in the world; what it needed was the punitive stick administered under effective safety laws.

CONCLUSION

The most important conclusion of this disasters study is that, had the law not been broken, most of the disasters described here would never have happened. On this we can do no better than let one of the Pittston disaster reports (Compass No. 2, 1963) have the last word:

> Evidence found during the investigation of this explosion as well as testimony offered at the official hearing showed clearly that in many instances men and officials in this mine, as has been true in nearly every other explosion, failed to follow known safe mining practices. The failure of supervisors and employees to follow and comply with known safe mining practices, company rules and regulations, and State and Federal laws was the basic cause of the explosion (p. 16).

Although this finding seems to justify a punitive law enforcement approach to regulation, it must be noted that in 38 percent of the disasters there was also significant uncertainty about the causes. This was even so for some of the disasters where a violation of law was known to have been responsible. For example, with the Appin disaster there were only two possible sources of ignition for the explosion—a fan that was illegally turned on when not in a flameproofed condition, or a lamp that was also in contravention of the regulations. Either way, a violation of law caused the ignition, but uncertainty about which source would have made it difficult to allocate blame for the disaster in a court of law.

The findings about the organizational defects that contributed to the disasters raise a number of questions to be taken up in ensuing chapters.

1. Given the frequency of inadequate plans as causes of disasters, is there a way of directing enforcement to ensure that company plans are both adequate and followed?

2. Given the frequency of communication breakdowns, can the responsibility to know and report dangers be enforced?

3. Given the regularity with which a disaster is the culmination of a pattern of inattention to safety, is there a way of focusing enforcement on the patterns of conduct, as opposed to specific regulations?

4. Given that inadequately drawn lines of responsibility were a cause of some disasters, can the law require organizations to delegate clear responsibility for each aspect of safety compliance?

It has been pointed out that many of the disasters were carbon copies of previous disasters. Many were similar in both the technical failures and the organizational failures that led to the catastrophes. This implies that coal mine disasters are not capricious acts of nature about which we can do little, but patterned, predictable events which are therefore eminently preventable.

# 3. A Look at Some Safety Leaders

The original plan for this study included visits to the five large U.S. coal-mining companies with the best safety records and the five with the worst records. It is perhaps not surprising that the five with the lowest accident rates all agreed to my holding discussions with their senior safety personnel; but the first three companies with high accident rates that I approached all declined to cooperate.

Consequently, the study was based on interviews with senior safety executives of the five companies that the U.S. President's Commission of Coal (1980, p. 42) found to have by far the lowest accident rates among the largest fifteen companies. Mine Safety and Health Administration statistics also show that these companies normally have the lowest accident rates among all companies, although comparisons are sometimes difficult because the accident rates of small companies can fluctuate wildly from quarter to quarter.

Interviews were also conducted with MSHA inspectors who had experience with these five companies' mines. And the final source of data was the companies' safety policies and safety program summaries, and other relevant company documents.

The five safety leaders serving as subjects are U.S. Steel, Bethlehem Steel, Consolidation Coal Company, Island Creek Coal Company, and Old Ben Coal Company. It would be a mistake to assume that simply because these companies at the time of writing have the best safety performance in the American industry, they are in every way exemplary in safety matters. Some mines owned by these companies have shocking safety records. Nevertheless, there must be some things we can learn about corporate compliance systems which keep the average lost-time accident rates of mines owned by these companies at about half, sometimes less than half, the industry average. Let us consider the safety compliance systems of each company in turn.

## U.S. Steel

United States Steel Corporation (U.S. Steel) was one of the companies recommended to me by officers of the Mine Safety and Health Administration as having an exemplary safety compliance program. The agency's accident statistics also support the conclusion that U.S. Steel is an industry leader in safety. Their lost-time accident rate for underground coal mines is about half the national average. The President's Commission on Coal (1980, p. 42) found U.S. Steel to have the lowest disabling injury incidence rate (lost workday cases per 200,000 work hours) in the industry for 1978-9.

U.S. Steel has a corporation-wide safety program that applies to coal mining in essentially the same way as to other corporate operations. The corporate "Safety Program" document is unusual in its emphasis at several points that safety has a higher priority than production. For example:

It is doubtful that any company ever made significant safety progress just by being "interested in" or "concerned about" safety, as it is so often expressed. Rather, management—top management—must have strong convictions on the necessity for placing safety first, above all other business considerations (p. 4).

The U.S. Steel philosophy strongly emphasizes the notion that safety is a management responsibility. Because of this, the company's senior coal executives tend to disagree with those in the American coal industry who advocate changes to the Mine Safety and Health Act to impose individual liability on miners for violations. The predominant view among U.S. Steel managers is that liability for violations ought to be corporate, and responsibility for discipling individual miners also ought to be corporate.

Further, the U.S. Steel philosophy is that safety is a line management responsibility, that is responsibility does not lie with a specialized safety staff, but with each manager in the production line of command. At corporate headquarters in Pittsburgh, there is a corporate safety department and a chief inspector of mines. However, the chief inspector has neither staff nor power to order mine managers to do anything. While the chief inspector's role is only advisory, his suggestions have considerable informal clout because he answers directly to the president of the U.S. Steel Mining Company.

Even out in the field, U.S. Steel's safety staff is small compared to American coal-mining companies of comparable size. U.S. Steel has only thirty-five designated coal mine inspectors. The company owns twenty mines and eight preparation plants, which are grouped into six districts. Inspectors based at particular mine sites advise the mine manager on safety. They can recommend that a section be shut down when they believe it to be unsafe. However, the mine manager can veto such a recommendation. It is the manager who is accountable for safety, not the inspector. The inspector directly reports to the chief inspector for his district. However, the chief inspector is also unable to order a mine manager to do anything—his job is to advise the district's general superintendent on safety. Of course, if a mine manager overrules an inspector, the general-superintendent may take advice from his chief inspector to pull the manager into line. It follows that, if there is a safety breakdown, accountability for the failure is not normally imposed on the safety staff but runs up line management. An impressive feature of the U.S. Steel safety program is that it defines in quite clear terms the safety responsibilities of different levels of the management hierarchy. If a serious breach of safety standards occurs, there is not much room for anyone's disclaiming responsibility by saying either, "I understood it to be my boss's responsibility," or "I understood it to be my subordinate's responsibility." To understand how this line responsibility works from day to day, we must consider in turn the ten elements of U.S. Steel's safety program:

1. Safe job procedures
2. Basic training
3. Individual contacts
4. Safety observations
5. Employee records

6. Accident investigation
7. Awareness charts
8. Physical conditions and inspections
9. Activity reports
10. Safety audits

## Safe Job Procedures

U.S. Steel's internal rules tend to particularism rather than universalism. Rule-making authority is decentralized to the districts and different U.S. Steel districts may operate under somewhat different safety rules. Within districts, most of the rules are specific to particular jobs. For every job to be done in a mine, a set of "safe job procedures" are written, usually by the "group discussion" method. The foreman selects two or more of his workers who regularly perform the job and explains that their knowhow is needed to develop a safe job procedure. Together they analyze hazards and discuss ways of requiring the job to be done to avert these hazards. After this discussion, the foreman fills out a "safety analysis sheet," which systematically lists all known potential hazards associated with performing the job. The foreman then writes a recommended safe job procedure. Once the requirements in this procedure are either amended or accepted by senior management, deviations from the procedures can be punished by suspenion of the offending employee from work for a certain number of days or, in extreme cases, dismissal. Compared with other American companies, U.S. Steel has a reputation for firing workers who breach safety rules. Foremen convene periodic "review conferences" at which workers can suggest revisions of safe job procedures in response to changing conditions.

## Basic Training

Once safe job procedures have been written, foremen have the basic responsibility to ensure that their workers are instructed, step by step, in the approved method of doing the job. Department heads are responsible for developing training plans to ensure that foremen provide all workers with all required basic training.

*Individual Contacts*

Each foreman must make at least one individual contact each week with each employee under his supervision to continue his training and consolidate safety knowledge. With inexperienced workers, these contacts are usually "tell-show" checks whereby the worker is asked to explain what should and should not be done and why the approved procedure is the safest one.

*Safety Observations*

Foremen are required to make at least two planned safety observations of each employee each month. The safety observations are planned so that they cover systematically all job operations for which the employee has received instruction. In addition to the safety observations, which are planned and scheduled at the beginining of each week, foremen are expected to perform additional "impromptu observations" following chance recognition of unsafe practices. Whenever a foreman observes an unsafe condition or work method, whether in a planned or impromptu safety observation, he must correct it immediately and report the occurrence to higher management on a "supervisor's safety report." An example of such a report is shown in figure 1.

Note in figure 1 that a distinction is made between a violation and "violation—lack of instructions." There can be no violation unless a worker engages in a practice contrary to a safe job procedure, safety rule, or specific safety instruction in which he has been trained.

*Employee Records*

The foreman can tell whether a worker who deviates from a procedure or rule has been trained in it by looking at the employee's record. For every employee a record is maintained by his foreman, noting his safety history—basic training, safety contacts, planned safety observations, unsafe acts, violations, discipline, and injuries. When workers move from foreman to foreman, their records move with them, so a new foreman can discover at a glance what safety training a worker lacks for his new job, and individual contacts and safety observations can be planned with him accordingly.

**SUPERVISOR'S SAFETY OBSERVATION REPORT**

O   Planned Observation     Occupation *KEEPER*

☒   Impromptu Observation

Name *JOSEPH R. MYERS*     Check No. *20972*

What was observed? *I OBSERVED THIS MAN WATERING THE IRON RUNNER BEFORE THE 20 MIN WAITING PERIOD WAS OVER. (SAFE JOB PROCEDURE 123)*

Where observed? *#1 BF CAST HOUSE*

CHECK ONE:

O   Unsafe Condition       O   Unsafe Act

O   Violation—            ☒   Violation of Rules, Pro-
     Lack of Instructions              cedures or Instructions

Action Taken or Recommended *REINSTRUCTED MYERS AND GAVE HIM A FORMAL DISCIPLINE SLIP*

Foreman *R. J. Collins*     Date *4/30/71*

General Foreman *G. Jones*     Date *4/30/71*

Superintendent's Review and/or Action *Instruct Myers if he violates this procedure again, he will receive time off.*

Superintendent *Fred Smith*     Date *4/30/71*

*Figure 1.* Fictitious example of a U.S. Steel supervisor's safety observation report.

## Accident Investigation

Every injury is reported and investigated by the foreman, regardless of its severity. For minor injuries, a foreman's report may be all that is required. But as injuries become more severe, committee investigations involving more and more senior management take place. Whenever there is a fatality, the general superintendent of the district and his chief inspector immediately go to the site and shortly thereafter, the president of U.S. Steel Mining Company and his chief inspector, as well.

Together with a member of the legal staff from corporate headquarters, this forms the nucleus of the investigating committee. A senior executive claimed that the committee generally succeeds in allocating blame to individuals for the fatality, using the predetermined, published safety responsibilities of different levels of management as a guide. Interrogation by the commitee is reputedly tough; as one executive observed: "They're not a pleasant experience for a mine manager."

On accident reports, questions must be answered about the responsibility of the victim and other persons, as well as the work environment for the accident. Under each of these headings, the "underlying cause(s)" must be identified. Another important part of the report must indicate whether the accident was the result of a job procedure that is defective in its provisions for safety, a failure to instruct in the procedure, or a violaiton of the procedure. Further corrective action taken and recommended must be analyzed.

> In order to utilize the accident experience in each plant as a positive force for preventing similar accidents in all other plants, a rapid communication routine is used. Information on all fatalities, disabling injuries, major fires, near-serious accidents, and unusual occurrences is telephoned immediately to Corporation Safety Headquarters in Pittsburgh. A preliminary report is sent immediately...to all Corporation locations which can benefit from it.... Four days later, reports summarizing the complete fact-finding results are distributed.

## Awareness Charts

Awareness charts are tally forms on which injuries, violations, and unsafe acts are recorded, by occupation. Once these figures identify an area where greater safety control is needed, a further tally may be conducted to pinpoint the specific types of violation that are causing most of the injuries within a particular occupational group during a given time period.

## Physical Conditions and Inspections

The Engineering Department enforces safety standards and specifications for the design of equipment and facilities. Newly

installed or changed facilities are inspected and approved for
safety before they are released for operation or use. Also,
"specific responsibilities are established for periodic inspection,
and for prompt correction of deficiencies or immediate shut-
down of equipment if a serious hazard is found" (p. 20). The
Engineering Department is notified of all accidents involving
equipment failures or deficiencies.

*Activity Reports*

Foremen, departments, and entire plants must all produce
summary safety activity reports either weekly or monthly. These
indicate how many safety contacts, observations, injuries,
disciplinary actions, job safety analysis conferences, unsafe
conditions, and inspections there have been during each week.
These reports ensure the accountability of foremen, department
heads, and superintendents for the safety performance of their
units.

The accountability mechanism for general superintendents
of mining districts is more interesting. The general superintendents attend a monthly meeting with the president of the
mining company and other senior executives, at corporate
headquarters. Each general superintendent, in turn, makes a
presentation on his district's performance during the previous
month—first, on safety performance (i.e., accident rates) and,
second, on productive performance (tons of coal mined). After
the safety presentation, the corporate chief inspector of mines
has the first opportunity to ask questions. If the accident rate has
worsened in comparison to previous months, or to other districts, the question invariably asked is, Why? I was told that
the twenty-four or twenty-five senior people who attend these
meetings exert a powerful peer-group pressure on general
superintendents whose safety performance is poor. It is an
extreme embarassment for general superintendents to have to
come back month after month and report safety performances
falling behind those of other districts.

These meetings, incidentally, also fulfill the function of
regulatory innovation. It was said earlier that each mining
district, rather than the corporation as a whole, writes its own
rule book. General superintendents who have introduced new

rules or technologies that have worked well in reducing accidents will score points by mentioning these successes in their reports. Other districts will then adopt these controls. An advantage of the combination of decentralized rule making and centralized performance assessment is that creative approaches to reducing accidents may be more likely to emerge than under the stultifying influence of a corporate book of rules.

*Safety Audits*

While the activity reports focus on the quantity of safety controls, safety audits focus on their quality. The general superintendent of each mining district must personally participate in an annual safety audit of at least one of his mines and facilities. He relies heavily on his chief inspector and other staff in these audits. However, his personal participation emphasizes the fact that it is the line managers, rather than the safety staff, who are ultimately accountable for safety.

*Summary*

The essentials of the U.S. Steel safety program are summarized schematically in figure 2. The main features of U.S. Steel's approach to safety are:

1. Formal and repetitive recognition in company policy that safety must be an integral part of production and even take precedence over it.

2. Decentralized safety rule making, rather than imposition of detailed corporate rules.

3. Great emphasis on specific rules defining how each worker must do his particular job, as opposed to general rules applying to all workers.

4. Routine use of peer group pressure on senior managers at meetings so structured that managers with poor safety performance suffer embarassment. Thus, decentralized rule making is combined with centralized performance assessment. There is clearly defined accountability for

safety breakdowns with ultimate responsibility resting with line rather than staff

5. Great emphasis on record keeping to ensure: that supervisory personnel maintain a vigorous campaign of safety training contacts; that increases in accident frequency are quickly noticed; that safety problems are rapidly communicated to those who can act on them; and that a plan of attack exists for dealing with all identified hazards.

6. Once employees have been properly trained, a relatively punitive approach (in comparison with other companies) to employees who violate safety rules.

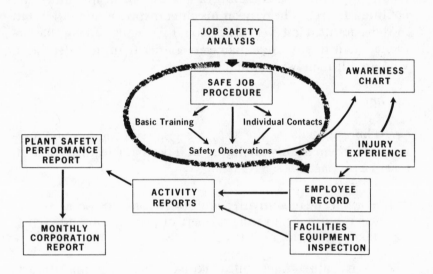

*Figure 2.* Essentials of the U.S. Steel safety program.

## BETHLEHEM STEEL

Bethlehem Steel shares a reputation for coal-mining safety similar to that of U.S. Steel. The statistical record shows Bethlehem to have about half the accident rate of the industry as a whole; indeed, if anything, in recent years Bethlehem has the superior coal mine safety performance of the two steelmaking giants. There are many respects in which the safety programs of the two companies are almost identical. Bethlehem has programs for writing safe job procedures, basic training, individual contacts, safety observations, employee records, accident investigation, awareness charts, inspection, and audit—all of which are in so many respects similar to those of U.S. Steel that the details will not be repeated here.

Control over these programs is even more decentralized at Bethlehem than at U.S. Steel. There are no detailed corporate rules. For example, corporate headquarters does not determine the frequency of safety contacts or observations. It is up to each mine to set its own standards on such matters. Certainly, corporate policy requires that all mines have programs for writing safe job procedures, safety contacts, equipment testing, and so on. Moreover, headquarters safety staff do receive copies of these plans and pass down criticisms of them. However, responsibility for rule writing and rule observance is fundamentally in the hands of mine managers. Whereas at U.S. Steel investigations following major accidents are basically run by corporate headquarters, investigations at Bethlehem are under the control of local operating management; corporate safety staff do attend these investigations and assure that adequate countermeasures are taken, but they do not run them.

Similarly, safety audits conducted by the corporate safety staff are viewed as every bit as much a service provided to line management by the head office as they are a mechanism for ensuring that mines are complying with corporate standards. There are six coal-mining divisions in different regions of the country. One mine in each division is audited annually by the corporate safety department. Often these mines are selected on the basis of requests by division management that their safety program at a certain mine would benefit from outside advice. Equally, however, there are times when audits are scheduled by corporate headquarters because they perceive a deficiency in a particular mine.

As does U.S. Steel, Bethlehem believes that "safety starts at the top." Corporate policy makes an effort to define the safety responsibilities of different levels of management, though not in the same detail as U.S. Steel. Under this definition of responsibilities, operating heads (mine managers) are primarily responsible for their operations' compliance with federal, state, local, and corporate safety codes and standards. As at U.S. Steel, safety is a line responsibility.

Each mine has a resident safety inspector who answers to the divisional (coal-mining region) manager of environmental safety and health. The latter, in turn, answers to the division's general manager. Neither the local inspector nor the manager of environmental safety and health is accountable for a safety breakdown. Inspectors cannot insist that mine superintendents do anything unless there is an imminent danger; their role is basically limited to advice. When the advice goes unheeded, the inspector can complain to his boss, but the manager of environmental safety and health is equally unauthorized to direct a mine superintendent. He can only implore his boss, the division manager, to pull the mine superintendent into line. The only difference from the U.S. Steel model is that the local inspector does not answer to the mine superintendent. Safety staff answers to line at the regional level, instead of the mine level. There is a larger corporate safety staff at Bethlehem. About ten persons at corporate headquarters could be counted as corporate safety staff across Bethlehem's total operations. These staff members are not concerned with "catching" safety violators, but with assisting with planning safety within working environments and reminding line managers of their safety responsibilities.

Bethlehem has a less punitive approach for dealing with violations of safety rules by employees. Only two or possibly three coal miners could be said to have been dismissed during the past seven years for safety violations. Suspending employees from work for a couple of days for safety violations is also rare; perhaps one or two coal miners a year would be punished in this way over safety matters. There is, however, a lot of moving people sideways when they do not perform a task well, either in terms of safety or on some other performance criterion. People are not fired frequently, because Bethlehem believes that it is possible to find a niche somewhere that will suit the talents of any individual.

At all levels, the emphasis is on informal social control to achieve safe work performance. Training for supervisors emphasizes the importance of motivating workers to want safety for their own sakes. The principle is that a punitive approach to safety violations would undermine the attitude instilled through-out the company that safety is in everyone's interest, rather than something management enforces in its own interests. Bethlehem does not have the monthly criticism and self-criticism sessions with regional chiefs that is a feature of U.S. Steel's informal control. Instead, the senior vice president of operations, coal, telephones each of his six division managers every day to get both their production figures for the day and their reports of any serious accidents. Also, monthly statistical reports of safety performance are required. Annual goals are set for each division; these include accident reduction goals. In an informal sense, failure to achieve these goals can affect promotion prospects, and there is said to be much heat from multiple sources when managers fail to achieve their targets. There is no formal mechanism for paying higher bonuses to managers who achieve better accident statistics than others.

Finally, as does U.S. Steel, Bethlehem manifests a strong commitment to safety in its company philosophy. However, its method differs. While U.S. Steel's emphasis is on assurances that safety is a higer priority than production, Bethlehem's phil-osophy is that there is no incompatibility between safety and production.

> Bethlehem policy is that safety is an integral part of production. The two cannot be separated. All accidents, regardless of consequences, are symptoms of production inefficiency. Efficient production is accident-free. Good safety and good production go together (*Bethlehem Steel Accident Prevention Program*, p. 3).

Bethlehem management believes that the real cost of a coal mine fatality is enormous in terms of compensation payouts, blows to miner morale, and the interruption of production. To reinforce the philosophy that safety and profitability go hand in hand, Bethlehem's accounting system charges compensation payouts and fines imposed for violations of safety laws to the mines concerned.

## Summary

Bethlehem has mechanisms essentially similar to those of U.S. Steel for ensuring that safety training and supervision are adequate, that safety problems are quickly communicated to persons who can act on them, and that a plan of attack exists for dealing with all identified hazards. These include programs for writing safe job procedures, basic training, individual contacts, safety observations, employee records, accident investigation, awareness charts, inspection, and safety audit.

Control over these programs is perhaps even more decentralized at Bethlehem than at U.S. Steel, while safety performance assessment is equally centralized and accountability equally imposed on line rather than staff.

The main difference between the two is that Bethlehem is less punitive in dealing with those who violate safety standards. The corporate philosophy is to depend less on command and control to achieve safety and more on bringing employees to understand that following safety rules is in the interests of both productivity and accident prevention, of both management and workers. The tone of safety compliance efforts is therefore less punitive and adversarial and more cooperative and flexible.

## Consolidation Coal Company

The Consolidation Coal Company is the second largest coal producer in the United States. It was taken over by Continental Oil (later renamed Conoco) in 1966, and Conoco, in turn, was acquired by DuPont in 1981. Consol's safety performance was the subject of a concerted public attack run by one of Ralph Nader's affiliates (Campaign Continental). Campaign Continental distributed literature to shareholders arguing, among other things, that "over a four year period from 1968 to 1972, Consol maintained the highest death rate per million man hours of any coal company, with 165 men dead." In each of the three years following Campaign Continental's attack in 1972, there was a drop in Consol's fatality rate, and since then it has remained well below the industry's average. One aspect of Campaign Continental was an unfavorable comparison of Consol's with Bethlehem's accident rate. The gap between Consol and Bethlehem narrowed during the 1970s to such a degree that Consol now deserves to be classified as one of the leaders in safety performance in the U.S. coal industry.

Consol's leadership is most clearly manifested in the size of its safety-related staff, which tripled in the ten years following Campaign Continental to 300 (with subsequent mine closures in 1983, the number dropped to 225). In absolute terms, Consol certainly has the largest safety staff in the industry and probably the largest proportion of its employees working on safety. In 1973, the year after Campaign Continental, Consol commenced a safety action program that drew heavily on U.S. Steel's program. It included the analysis of safe job procedures, maintenance of individual employee safety records, safety awareness charts, on-site investigations of serious accidents conducted by senior head office staff, and annual audits by safety staff—all of which we saw were features of the U.S. Steel system. Basic training of newly employed apprentice miners also received heavy emphasis. Consol's expenditure on training per employee increased by 325 percent in that year. Corporate policy requires new miners to be at all times within not only the sight but also the sound of an experienced miner during their first ninety days on the job; and within either sight or sound for the first six months. Consol is also closer to U.S. Steel's more punitive approach to miners who break safety rules than to Bethlehem's nonpunitive approach. Sanctions of five or ten days off without pay are not uncommon at Consol. When company inspectors detect unsafe practices, there is the same emphasis as in the other two companies on ensuring, first, that dates are set by which remedial measures must be implemented, and second, that a record is made of whether the remedy is implemented by that time.

Consol's five regions and forty mines are encouraged to compete with each other for good accident rate performance. The percentage of corporate safety violations accounted for by each region is also a regular part of company statistics. There are not the structured monthly criticism and praise sessions for the safety performance of senior managers that there are at U.S. Steel. However, at very frequent operating meetings between the executive vice president, operations, and the five senior vice presidents of mining, safety is always first on the agenda, and vice presidents are rapped on the knuckles when accident rates are not declining in their mines.

The organization of safety within the company is quite different from the two steel giants, however. Essentially, there

are two types of safety staff—those who report to line managers
and those who report through staff channels to the vice
president, safety. Most mines, and all the company's large mines,
have a supervisor of safety who reports to the mine superintendent. Each of the five regions has a regional safety manager
who reports to the senior vice president of mining in charge of
the region. In addition to this solid line reporting relationship,
there is a dotted line from the regional safety manager to the vice
president, safety.

The second set of safety personnel reports entirely through
staff channels. Each region has a chief inspector, who is
responsible for a number of mines and reports to the corporate
safety director. The safety director reports to the vice president,
safety, who, in turn, reports directly to the executive vice
president, operations. This inspectorial staff does not have even
dotted line linkage to the operating managers at their worksites.
In principle, it should therefore be impossible for a mine
superintendent to instruct his safety inspector not to report to
corporate headquarters any serious safety deficiency at the mine.

*Summary*

Consol has the same basic program as the two steelmakers
for ensuring that safety training and supervision are adequate,
that safety problems are quickly communicated to persons who
can act on them, and that a plan of attack exists for dealing with
all identified hazards. Consol also relies on similar competition
between regions and mines on statistical safety performance.
However, this takes place in the context of a far more centralized
corporate approach to safety and with a much larger safety staff.
Line accountability is still the fundamental precept. But, in
addition to safety personnel who report to and advise line
managers, there is a seond set of safety personnel who report
solely through staff channels to the vice president, safety.

ISLAND CREEK COAL COMPANY

At the time of the establishment of the President's Commission on Coal (1980), Island Creek was ranked as the third safest

major coal-mining company in the United States. Since 1978-9 its lost-time accident rate has held at about the same level. Island Creek has many of the features we have seen among the other American safety leaders: analysis of safe job procedures; more stringent training requirements than are mandated by law; a requirement that foremen make a meaningful weekly safety contact with all the workers under them; involvement of senior corporate officers (e.g., the vice president, safety) in serious accident investigations; detailed employee safety records; and well-organized weekly safety meetings of fifteen minutes' duration with the miners. Island Creek is unlike U.S. Steel, but like the other three safety leaders, in its willingness to dismiss employees for safety violations only in extraordinary circumstances. The company has no regular program of corporate safety audits.

In one area, there is a difference of emphasis in Island Creek's safety compliance program: the company uses tangible incentives for safety performance. Section foremen receive some extra pay for every month in which their sections run without a lost-time injury. The amount involved is not large, varying up to a maximum of $50 for the month. Foremen also receive small bonuses for maintaining an established ratio of inspections to violations for governmental inspector visits to their sections. Each year the corporation sets an accident rate target for its mines that is rather difficult to obtain. Every employee at those mines that achieve the target receives a special small gift; in some mines a gift catalogue is circulated and employees choose their own gift, worth about nine dollars.

Island Creek has two company safety inspectors at most of its twenty underground mines. Some smaller mines have only one inspector. Where there are two inspectors, one is primarily devoted to dust and noise compliance. Island Creek has four coal-mining divisions: two in Kentucky, one in Virginia, and one in West Virginia. Each division has a safety director to whom the mine-based inspectors have a dotted line reporting relationship; inspectors also have a rather firmer reporting line to their local mine superintendents. The divisional safety director, in turn, has a solid line to his division president and only a dotted line to the corporate vice president, safety.

Corporate headquarters monitors divisional safety performance at monthly meetings, when the division presidents visit the head office and comparative safety performance is reviewed.

Island Creek is a subsidiary of Occidental Petroleum, a corporation not famous for either its occupational safety or its environmental record, particularly as a result of another of its subsidiaries' responsibility for the Love Canal fiasco. However, Occidental's involvement in Island Creek's safety program is minimal. Island Creek's vice president, safety, does have a dotted line reporting relationship to Occidental's Safety Department, as well as a solid line to the executive vice president of Island Creek. However, Occidental imposes no special safety policies on Island Creek. The identification of employees is clearly with Island Creek, rather than Occidental, as their corporate employer. Indeed, a great many Island Creek miners are not even aware that their company is owned by Occidental. Consequently, the poor national reputation of Occidental has not created a problem for Island Creek in building employee respect for its commitment to safety.

*Summary*

In common with U.S. Steel, Bethlehem, and Consol, Island Creek has a range of formal programs to guarantee that safety is not neglected: analysis of safe job procedures, intensive training, safety contacts and meetings, employee safety records, etc. There is also the same emphasis on line accountability and decentralized control over safety, combined with centralized evaluation of safety performance. One difference from the other three large companies is the use Island Creek makes of financial incentives to motivate safety compliance.

OLD BEN COAL COMPANY

Of the large American coal miners, the company with the lowest lost-time accident rate in 1980 and 1981 was the Old Ben Coal Company. But Old Ben was not always the safest coal-mining company in the U.S. Along with Consol, Old Ben has a reputation with the MSHA officials with whom I spoke, in both Washington and southern Illinois (where most of the company's mining takes place), for having dramatically improved its safety performance in recent years. Some MSHA officials attributed this to a new emphasis on safety that was gradually imposed in the years after Old Ben was taken over by Sohio, in 1968. Compared with the other leaders in safety we have discussed, Old Ben is relatively small, employing 2,700 miners at 5 mines.

How can we explain the dramatic success of Old Ben in so improving its accident rates as to become number one in the early 1980s? Certainly not by increases in the safety staff, as seemed to be the case with Consol. Between 1978 and 1981, the company's safety and dust-counting staff was reduced from fifty-two to thirty-two. Some people within the company, and at MSHA, attribute the improvement in recent years to the shift of Elmer Layne from Island Creek, the company which formerly had the best accident rate in the industry, to manager of corporate safety at Old Ben, in 1978. Although one should be suspicious of explanations for organizational change based on the personality of a single person, Elmer Layne's peculiarly Southern charisma does make such an explanation tempting in this case. He commands enormous respect from his safety staff. I sat in on one of their regular Tuesday night meetings, at a bar in soutern Illinois, at which problems were solved and esprit de corps consolidated over the odd beer. Another consequence of Mr. Layne's compelling personality is that the quality of corporate relationships with MSHA has changed. One is tempted to wonder whether quite a bit of the considerable drop in MSHA citations in recent years could be due to a more trusting attitude on the part of the agency toward an Old Ben under Mr. Layne's safety leadership. Certainly, during Mr. Layne's tenure, relationships with the agency have been less litigious than previously.

The success of Mr. Layne and his staff can be attributed to the fact that they have clout within the organization. When a company safety inspector orders a section of a mine shut down, corporate policy forbids the mine superintendent to overrule the decision.

Each mine has two inspectors and one training officer, who answer to the safety director for the mine. The safety director has a dotted line to the mine superintendent and a solid line to the manager of corporate safety. In spite of the fact that the safety staff does not really answer to the superintendent, it is primarily the superintendent who is held accountable for safety violations and accidents.

It probably is true that the Sohio takeover of Old Ben resulted in an improved attitude toward safety. However, Sohio corporate headquarters in Cleveland plays a relatively minor role in the day-to-day work of building safety. Sohio has a safety auditing staff, albeit one lacking in coal-mining expertise, and it audits one mine each year for compliance with coal mine safety laws.

Safety is controlled by officers in the field in southern Illinois and Indiana, from Mr. Layne down. At the time of my visit, there was no person responsible for safety in Old Ben's Chicago headquarters. There are no head office safety meetings, as at U.S. Steel, where superintendents have their safety performance scrutinized. If there is a fatality or a very serious accident, however, headquarters becomes involved; the second in command of the company (the executive vice president, operations) visits the site to participate in the investigation. This senior person also personally views and initials all MSHA citations. The company has been known to dismiss people for safety violations, but this is rare. Greater reliance is placed on social rewards than on punishments: a safety banquet is held each year at which top management presents safety awards. Mines and regional offices that achieve their safety targets fly a white flag with a green cross near their entrances.

The general picture of low head-office control over safety must be tempered in one important respect. All Old Ben executives, down to the level of mine superintendents, set targets for themselves each year. These targets include safety (e.g., reducing the lost-time accident rate to X). When decisions are made in Cleveland each year on the bonuses to be paid to Old Ben executives, considerable account is taken of whether they achieved their various targets.

Although the improved safety performance at Old Ben in recent years is undeniable, it must be pointed out that this has undoubtedly been made easier by favorable conditions at their mines, which are generally nongassy, in thick coal seams, with stable roofs, and an increasing use of longwall mining techniques that minimize the risks of roof falls.

*Summary*

Old Ben has manifested a dramatic improvement in safety performance in recent years to become an industry leader. The improvement certainly cannot be explained solely by an increase in safety staff, as at Consol. More likely the explanation lies with the increased clout given the manager of corporate safety within the organization, and perhaps also the personal qualities of the individual currently filling that role.

Old Ben does not have the detailed corporate safety policies of U.S. Steel, Bethlehem, or Consol. Nor does it have the array of formal programs for writing safe job procedures, individual contacts, safety observations, employee records, awareness charts, and the like. Nevertheless, one does get the impression that more informal mechanisms, fomented by an activist cadre of company inspectors with clout, achieve the same assurance of supervision, training, and communication of safety problems. In spite of the importance of the badgering of an independent and sometimes unpopular safety staff in achieving lower accident rates, ultimate responsibility for safety—as with the previous three safety leaders—lies with line rather than staff.

## Conclusion

*Differences among the Five Safety Programs*

"You can't cookbook safety," Bethlehem Steel's director of safety said to me during our interview. He was becoming a trifle annoyed with my constant questions about the place of safety within the organization—who answers to whom, and the like. The senior vice president for operations, coal, also felt my questions were misguided. He pointed out that even though Bethlehem was a leader in safety performance, there might be very little that other companies could learn from Bethlehem in terms of formal structures, because each company has a unique history, a unique set of personalities in senior positions, and different organization charts, and, consequently, each must find a unique solution to the problem of the place of safety within its structure.

I confess the criticism was apt. In these interviews I suppose I was searching for some magic formula that would be evident in all the companies with the very best safety records. Then, I thought, perhaps it would be possible to enact laws to require other companies to adopt this same formula.

One hunch was that the safety leaders would be companies that granted their inspectors independence by having them answer to a safety department rather than to the mine superintendent. The theory here was that safety would be less likely to be compromised when the inspector could only be overruled by another safety professional, rather than by a line manager whose primary concern was production. In fact, I found that, at U.S. Steel and Island Creek, inspectors at the mine, chief inspectors at the district level, and the senior safety person at the corporate level all reported directly to the line managers at their own levels. At Island Creek, safety staff at different levels of the organization had only a dotted line connection to each other. At the other extreme, Old Ben showed only a dotted line from inspector to mine superintendant, while solid lines connected the inspector to the director of safety and the director of safety to the manager, corporate safety. Consol had an unusual compromise, with one set of safety staff reporting to line managers and another set reporting through staff channels to a vice president, safety. Bethlehem had yet another sort of compromise, namely, mine inspectors with a solid line to the divisional manager of safety and health and only a dotted line to their mine superintendent, while the divisional manager of safety and health answered not to a corporate safety person, but to his divisional general manager.

In other words, here we have five companies, all of them safety leaders, and among them exists the whole range of conceivable reporting relationships for safety staff within the organizational power structure.

The companies also have quite different approaches to enforcing compliance with their safety rules. U.S. Steel's approach is rather punitive; employees are frequently dismissed or given days off without pay for failing to comply with safety standards. Consol also not infrequently adopts this punitive stance, whereas the other three positively reject such punitiveness in building motivation for safe practices among employees.

Island Creek is different from the others in the way it uses financial carrots rather than disciplinary sticks to encourage safety. Although none of the other companies make explicit payments to employees for achieving improved safety, in varying degrees they do incorporate safety performance into the overall evaluation of managers for promotions or bonuses.

The size of safety staffs is another variable in which the five companies differ. At one extreme is Consol with a safety staff that peaked at 300; at the other, U.S. Steel, with a staff of 35. On the one hand, Consol achieved a striking improvement in accident rates after trebling its safety staff; on the other, Old Ben achieved even more remarkable improvement by reducing and rationalizing its inspectorial force.

In summary, neither the place of safety in the formal power structure of the organization, nor the human resources dedicated to safety, nor the punitiveness with which safety is enforced, nor the use of tangible rewards for safety performance seem to be factors for which a common successful approach is evident among our five safety leaders.

## What the Five Companies Have in Common

Even though the place of safety departments in the formal organizational structure varies a good deal for all of the companies it is clear that safety personnel have considerable informal clout. Moreover, in all cases this derives from a corporate philosophy of commitment to safety and communication of the message that top management perceives cutting corners on safety to achieve production goals as not in the interests of the corporation. The way this is justified in company philosophy is quite different (witness U.S. Steel's philosophy that safety is a more important goal than production, and Bethlehem's philosophy that there is no incompatibility between the goals of safety and productivity); yet, in all five the effect seems to be an unwillingness of line managers to ignore the advice of safety staff. When a company inspector recommands closing a section of a mine because it is unsafe, in all of these companies line managers consider it inadvisable to ignore the recommendation because of the substantial risk that top management will back the safety staff rather than themselves.

In all five companies, the line manager, rather than the safety staff, is held accountable for the safety of his work force. Another universal feature is a clear definition of the safety level of the hierarchy that will be held accountable for different types of safety breakdowns. They are all companies that avoid the problem of diffused accountability—people know where the buck will stop for different kinds of failures.

Control over safety programs is also relatively decentralized in all five companies. This came as a surprise because some are renouned as highly centralized corporations. Pointing out how ironical it was that control over safety is so decentralized, one Bethlehem Steel executive said: "Bethlehem is probably close to the most centralized corporation in the United States." Decentralized control over safety is also perplexing because it is conventional wisdom that the coal mines owned by large, highly centralized steel companies have low accident rates because they have imposed the rigorous approach to safety of their steel operations on their mining investments.

However, while all these companies have decentralized control over safety, they also all have centralized assessment of the safety performance of line managers. All carefully monitor each mine and each district to ascertain whether their accident and fatality rates are improving or worsening in relation to the performance of previous years and to the performance of other mines and districts. Again, this centralized monitoring of performance is achieved in different ways by different companies—from the monthly criticism and self-criticism sessions at U.S. Steel, to the routine daily telephone calls from the senior vice president, operations, at Bethlehem, to the safety targets set for Old Ben executives by Sohio. But, for all of them the sense that the head office is watching their safety performance is pervasive.

Four of the five corporations have a set of programs that build in guarantees that safety training/supervision, and communication and rectification of safety problems, are working as they should. These include the formal requirements for writing safe job procedures, basic training, individual contacts, employee records, accident investigation, and audit (Island Creek being an exception to the latter), which were described in greater detail in the U.S. Steel case study.

The one company without these formal programs to ensure that safety is being given the importance it requires is the smallest one, Old Ben. It might be that smaller companies can achieve great success at minimizing accidents simply by hiring a charismatic manager of corporate safety who enjoys the backing of top management, whereas larger companies must depend on more formal organizational guarantees. In Weberian terms, larger corporations cannot rely on charismatic leadership to achieve their goals, at least not in the long term, and must opt for some sort of routinization of charisma.

In the final analysis, the conclusions about what these five companies have in common could be regarded as mundane. They are companies that:

1. Give a lot of informal clout to top management backing to their safety inspectors.

2. Make sure that clearly defined accountability for safety performance is imposed on line managers.

3. Monitor safety performance carefully and let managers know when it is not up to standard.

4. Have mostly formal (though informal in the case of Old Ben) programs for ensuring:

   (a) that safety training and supervision (by foremen in particular) is never neglected;

   (b) that safety problems are quickly communicated to those who can act on them;

   (c) that a plan of attack exists for dealing with all identified hazards.

These conclusions may seem banal; the industry's old-timers will react to them by saying there was no need to waste time on all that research to discover what they had always known. On the other hand, we can all feel more comfortable with conventional wisdom when there is some empirical research to support it. and we must remember also that, in the previous section, some other items of conventional wisdom about what distinguishes safe companies were challenged.

The other important point about these conclusions is that they mirror so closely those from the study of disasters. Poor communication of hazards, sloppy supervision, lack of plans to deal with forseeable hazards, inadequate training, unclear lines of responsibility—all were identified as common causes of mining disasters. And these are the very things we have found our industry leaders in safety to have under control. We see a convergence of findings from the two studies on what is necessary to save lives in coal mines. The cynical reader may wonder if the author having read the findings of the disaster reports, then imposed this reality on his interviews with executives from the leading safety companies. They would be wrong; the interviews were conducted, and this chapter was written (apart from the present conclusion), before I had read any of the disaster reports except that on the Appin disaster.

## Comparison with Results of Similar Studies

The foregoing research was also conducted without knowledge of two similar studies by MSHA—one by the agency (then the Bureau of Mines) in 1963, and another that had not been completed at the time of my fieldwork. Again, the findings were remarkably consistent with those of the present chapter.

The first of these studies (Davis and Stahl 1967) was based on interviews to analyze the safety organizations and activities of twelve companies that had won National Safety Council or Joseph A. Holmes Safety Association awards between 1957 and 1961. All these companies had at least one mine with zero lost-time injuries during a single year. Seven features of the safety organizations or activities were found to be common to all twelve companies:

1. A sincere desire on the part of both management and employees to prevent injuries.

2. A very obvious effort to determine why certain incidents occur and then to do everything possible to eliminate the hazards or actions which lead to injuries.

3. An obvious pride in safety accomplishment as well as mine performance.

4. A well-organized, active safety department respected by management and employees alike.

5. An eagerness to discuss safety matters with anyone who might contribute something to further improve their efforts.

6. Acceptance by the frontline supervisor of his responsibility for the prevention of injuries with assistance by the safety department and top supervision in his efforts.

7. A safety director with staff status or one responsible to an official charged with the overall operation of a group of mines (Davis and Stahl 1967, pp. 1-2)

Additional positive features were used by some, but not all, of the twelve companies:

1. Safety-hazard analysis preliminary to the formulation of safe work procedure or safety rules.

2. Training of all personnel—both supervisors and workmen—in the accepted safe job procedure.

3. Followup on training in safe job procedure.

4. Daily safety contact of foremen with workmen where some facet of the job procedure is mentioned to each workman daily and a record is made.

5. Periodic examination of both foremen and workmen in safe job procedure.

6. Attitude testing of safe work procedure violators and accident victims in a spirit of helping the victims rather than criticizing them.

7. Investigation of near accidents and requirement that the job procedure be restudied or that an analysis be made if none has been done previously.

8. Photographing hazards or hazardous procedures for use at safety meetings.

9. Safety letter to families of workers enlisting family support in promoting safe work habits.

10. Use of radio and television to promote safety.

11. Planned safety programs for a year in advance.

12. Complete accident analysis so that weak points in the safety program may receive immediate attention.

13. Orientation program for new workers.

14. Short daily safety meetings of sections, groups, or small units (Davis and Stahl 1967, p. 26).

Without going through the two lists point by point, it is clear that most of the safety organizations or activities found to be common in low-accident companies can be subsumed under the four features summarized from the present study in the previous section.

The second MSHA study (DeMichiei et al. 1982) compared twenty-one underground coal mines having exceptionally high accident rates with nineteen matched mines having exceptionally low accident rates. The comparison was based on three data sources: direct observation of mining by two researchers for about a week at each mine, interviewing miners, and questionnaires to miners, management, and MSHA inspectors.

Mines with greater entry heights and smaller numbers of sections were found to be safer. There was a variety of findings about the association between good training programs and safety. However, the important findings for our purposes related to organizational differences between mines with high and low accident rates:

Management/labor relations tend to have a positive impact upon a mine's accident and injury experience when:

Both management and labor have a positive attitude toward safety and health;

Open lines of communication permit management and labor to jointly reconcile problems affecting safety and health;

Representatives of labor become actively involved in issues concerning safety, health and production; and Management and labor identify and accept their joint responsibility for correcting unsafe conditions and practices.

Safety and health conditions are improved when:

Standard operating procedures are established, understood, and implemented;

Management equitably enforces established policies concerning absenteeism, job assignments, and standard operating procedures;

Formal safety and health programs are communicated to all employees and subsequently implemented by management and labor; and

Safety department has top management support in terms of funds, manpower, and the authority necessary to implement the safety and health program.

To keep abreast of all problems related to safety and health mine management is actively involved on all shifts.

Mine plans are thoroughly reviewed by management, labor, and MSHA to ensure that such plans incorporate measures necessary to adequately control the physical environment of a coal mine (DeMichiei et al. 1982, p. i-ii).

The themes are familiar—good communications, clear accountability, quality training, sound standard operating procedures, clout for the safety department, and good planning. As in the analysis of disasters in chapter 2, good planning was found to be especially important. The mines with high and low incidence rates were found not to differ in the adversity of the mining conditions encountered. However, there were striking differences in the ways by which mines planned to control geological adversity:

> ...at seven low incidence rate mines that were experiencing adverse roof conditions such as unconsolidated shale, clay veins, and extreme pressures, management had revised the existing roof control plan to incorporate roof control measures far exceeding minimum requirements. Furthermore, management had taken the necessary steps to ensure that employees responsible for implementing the revised plan were familiar with the approved roof control plan and the functions of the support being used.
> Of the 21 high incidence rate mines, nine encountered adverse roof conditions similar to those mentioned above. Observations revealed that only three of the nine mines had taken additional measures to control the adverse roof conditions. The other six had not taken the necessary steps to install additional roof support.... Indicative of this problem was the number of citations issued for violations of roof control standards at these six high incidence rate mines (DeMichiei et al. 1982, pp. 13-14).

A related interesting finding was that poor planning not only directly affected safety, but may also have done so indirectly by lowering morale:

> ...in seven high incidence rate mines, miners stated that poor morale was attributed to management's inability to plan effectively. Many times job assignments were conflicting, tools and materials necessary were not available, and management was often not receptive to the miners' concerns (DeMichiei et al. 1982, p. 19).

Another finding—not in the summary quoted above, but nevertheless important—related to the failure of mines with high accident rates to provide clearly established line accountability for safety:

> ...in five high incidence rate mines, the mine superintendents had no direct involvement in the mine's safety and health program. Responsibility for implementing the program was mainly the safety department's (DeMichiei et al. 1982, p. 20).

In short, there is a considerable convergence among the analysis of disasters, the analysis of safety organization in the five leading companies, and the two MSHA studies, concerning what organizational characteristics make for safe mining. These are clout for the safety department, clearly defined accountability for safety imposed on line managers, top management commitment to and monitoring of safety performance, programs for guaranteeing safety training/supervision, effective communication and, most important of all, effective plans to cope with hazards.

# Part II
# To Punish or Persuade?

# 4. Punishment and Persuasion: What Works?

This chapter first assesses the case for punishment as an effective regulatory strategy, then the case for persuasion.

## PUNISHMENT

To begin, consideration will be given to the proposition that punishing offenders for violations of mine safety laws prevents death and injury. The proposition can be broken down into three hypotheses that must be examined before one accepts the view that punishment will save life and limb:

1. Violations of safety laws cause accidents.
2. Inspections prevent violations.
3. It is the punitive aspects of the inspector's role that are responsible for preventing the violations and the accidents.

### Do Violations Cause Accidents?

It is beyond dispute that, if there were 100 percent compliance with mine safety laws, the majority of miners who die in coal mines would be saved. It is simply not the case that most fatalities arise from individual human error in circumstances where every effort has been made to comply with the law. In thirty-three of the thirty-nine disaster reports analyzed in chapter 3, serious safety violations were discovered in the course of investigating the accidents. For twenty-five of these cases, it was found that such serious violations either caused the disaster, were among the causes, or made the disaster worse than it would have been without the violations. For only two of the disasters was it found that the company concerned had a basically good compliance system, no improvement of which could have prevented the human error or forces of nature from causing the disaster.

Another study (McAteer 1981, p. 942) reported an analysis of all 126 Mine Enforcement and Safety Administration reports on *nondisaster* fatalities in 1975. Violations of mandatory safety standards were found by MESA to be at least a contributing factor in 72 percent of the fatal accidents. For roof fall fatalities, the figure was higher, 83 percent (36 out of 43 roof fall fatalities were at least partially a result of law violations).

The less serious the accident, probably the lower the likelihood that a violation of law was responsible. For example, a large proportion of minor injuries are caused by individuals slipping and falling, and except insofar as management has failed to comply with illumination regulations or programs to remove obstacles from mines, such accidents are not normally attributable to law violations. There is little systematic data on the proportion of nonfatal injuries caused by safety offenses. The only available evidence was collected by Zelonka (1974): a panel of coal-mining experts rated 196 U.S. accident, injury, and illness reports. About 21 percent of injuries were related to noncompliance with safety regulations, the percentage increasing as the accidents became more serious. It seems safe to conclude that, while the majority of fatalities could be prevented by law observance, probably the picture for nonfatal coal mine injuries is similar to nonfatal injuries in other workplaces, where U.S. studies have variously estimated between 10 percent and 30 percent of injuries are caused by violations of the Occupational Safety and Health Act (Mendeloff 1979, pp. 86-87; for some British data, see Veljanovski [1983, p. 88]).

Industry's critics of regulation make much of the fact that, although three-quarters of citations for violations fall into four major categories (electricity, ventilation, combustible materials, and fire protection), only 6 percent of nonfatal accidents in the United States between 1970 and 1977 were related to violations in these four categories (Consolidation Coal Company 1980, p. 41). It is true that most enforcement efforts are devoted to violations likely to cause fatilities, and more particularly, major disasters (e.g., unsafe electrical equipment sparking an explosion). This is because the unsafe environmental conditions that cause fatalities are more deterrable, being under management's control, than are the behavioral deficiencies of individuals who cause so many minor accidents. (We will argue later that deterrence is a more viable doctrine when it is directed at changing corporate environments than when directed at bending individual psychology.)

What is more important, the critique is unreasonable because it uses the success of regulation as an argument against regulation. Earlier in the century, the proportion of injuries from explosions and fires was far, far higher than today (Royal Commission on Safety in Coal Mines 1938, pp. 306-7; Bryan 1975, p. 67; Turton 1981). A major explanation for so low a proportion today is the effect of regulation. If enforcement of the four most frequent types of violations eased off, we might observe a reversion to the proportion of accidents attributable to these offenses of bygone years. It is like saying that electronic surveillance at airports is a waste of money because so few planes are hijacked, when we know that hijacking only began to decrease after the introduction of electronic surveillance and associated controls.

## Do Inspections Prevent Violations?

In gross terms we can see an historical association between the increased use of coal mine safety inspectors and declining fatality rates. Great Britain, France, and the United States have all progressively expanded their investment in coal mine inspection over the last 100 years. In Britain, the fatality rate per 1,000 coal miners today is about a twentieth of what it was in the 1850s (Collinson 1976, p. 3.2). In France, the decline was less dramatic because France led the world in coal mine safety regulation in the mid-nineteenth century; the death rate in 1860 was only five times the level of 1960 (Dardalhon 1964, p. 46). However, the improvement is tenfold for the same period if we look at fatalities per million tons of coal instead of per 1,000 miners (Dardalhon 1964, p. 47).

Data at hand for the United States do not go back so far, but in the last 50 years there has been about a fivefold improvement in fatality rates per 200,000 employee hours (a more conservative index of improvement, given the shortening of the work week) and about a tenfold improvement in fatalities per million tons of coal (National Academy of Sciences 1982, p. 39). Over this last 50 years the reduction in the fatality rate has been far greater than in Britain and France, even though the U.S. fatality rate per million employee hours is still three times that of Britain (see also National Academy of Sciences [1982, pp. 42-43]). The steeper improvement in the U.S. is undoubtedly the result of its having commenced the period with a more laissez-faire approach to safety regulation, but perhaps also because the United States has seen the greatest proportionate increase in its work force of government inspectors over the past 50 years. While federal coal mine inspectors increased from zero in 1930, to about 1,400 in 1980, the British inspectorate actually declined from 127 in 1938, to 95 in 1980. This is not a very meaningful comparison, however, because while the American coal industry was expanding during this period, the British industry was contracting, and with nationalization came the appointment of hundreds of de facto additional government inspectors by the National Coal Board itself.

The increase in resources devoted to coal mine safety regulation in the United States has not been steady over the last 50 years, but something that has happened in fits and starts. As in other countries, there have been several historical turning points at which major disasters shocked the public to the point of demanding tighter regulation of the industry. The three major federal legislative initiatives since the depression were all born of horrifying disasters: the 1941 Mine Inspection Act followed the deaths of 91 miners in Bartley, West Virginia, on 10 January 1940, and the loss of 73 lives at Neffs, Ohio, on 16 March 1940. In 1951, a series of explosions, including the West Frankfort, Illinois, blast that took 119 lives, ushered in the 1952 Federal Coal Mine Safety Act. The 1969 Federal Coal Mine Safety and Health Act was primarily a response to the explosion at Consol's Farmington, West Virginia, mine, in which 78 perished.

Lewis-Beck and Alford (1980) have undertaken a multiple interrupted time-series analysis to examine the impact of these three legislative initiatives on fatality rates per million employee hours. It can be seen from the estimated regression lines in figure 3 that the results were quite clear. During the 1930s, before the federal government decided to move in and supplement state enforcement of mine safety, fatality rates were on a plateau.

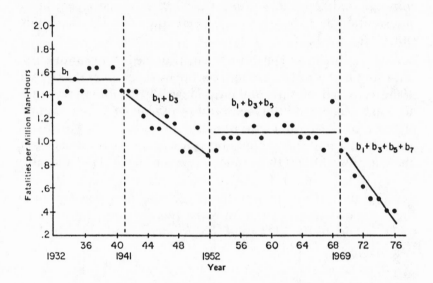

*Figure 3*. Multiple interrupted time-series analysis of the U.S. coal mine fatality rate (1932-76). Reprinted with permission from Lewis-Beck and Alford (1980:749).

Passage of the 1941 law, which for the first time granted federal inspectors the power to enter mines, was followed by a steep drop in fatality rates. Safety conditions in mines were entering a plateau when the 1952 legislation was passed, and the 1952 legislation was too weak to push the fatality rate below that plateau. Lewis-Beck and Alford argue that the 1952 act was a classic example of the "symbolic" law that radical critics of business regulation condemn. It did not result in tougher regulation, but merely served to placate the public demand that "something" be done. In contrast, the 1969 act was a tough law, indeed, perhaps the toughest regulatory statute one could find anywhere in the world. We will discuss later some of its provisions, such as mandatory citation of violations and a mandatory minimum of four inspections per mine per year. It can be seen from figure 3 that the reduction in fatalities following the 1969 act was about twice as steep as the reduction from the 1941 law. After Lewis-Beck and Alford entered controls for changes across time in technology, mine size, and type of mining (underground versus surface), the effects of legislative intervention described in figure 3 were found not to be spurious, but as fully robust.

Lewis-Beck and Alford then proceded to examine the reasons for the dramatic impacts on fatalities of the 1941 and 1969 acts and the failure of the 1952 law. They found that most of the variance in fatality rates could be explained by one factor— the size of the federal government's budget allocation to coal mine health and safety regulation. A critical difference between the effective acts and the symbolic one was that in 1942 and 1970,

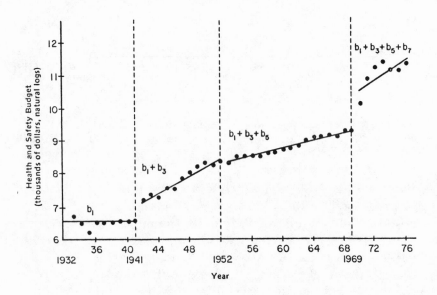

*Figure 4.* Multiple interrupted time-series of U.S. federal coal mine health and safety budget (1932-1976). Reprinted with permission from Lewis-Beck and Alford (1980:753).

following the 1941 and 1969 acts, there were sharp jumps in the budget allocation for coal mine health and safety regulation, while in 1953, following the 1952 act, there was not. It can be seen from figure 4 that 1941 and 1969 were turning points that increased the rate of growth of the health and safety budget, while 1952 was a turning point that slowed the rate of growth.

Overall, the correlation between size of the federal government's budget for health and safety regulation and fatality rate was -.78 (virtually unchanged at -.77 after deflation). However, the assumption of linearity in calculating the correlation coefficient is unsatisfactory because the budget variable follows an exponential path. Lewis-Beck and Alford therefore applied a natural logarithmic transformation to "straighten out" the health and safety budget. This transformation boosted the correlation with the fatality rate to -.90.

Since most of the health and safety budget is spent on the employment of inspectors and their support staff, the Lewis-Beck and Alford data warrant the conclusion that, when inspection increases, fatalities decline. Their data only cover the period until 1976. What has happened since can only have strengthened the correlation. The number of coal mine safety inspectors continued to grow until it reached a peak in 1978; this was the year in which the lowest number of fatalities until that time was recorded, 106. Fiscal belt tightening reduced the inspectorate after 1979, particularly in the early years of the Reagan Administration, and fatalities jumped substantially until a new record low was achieved in 1983. In the first year of the Reagan Administration staffing of the inspectorate dropped by 8 percent and fatalities rose 14 percent, even though 10 percent fewer hours were worked in 1981 because of strikes.

Lewis-Beck and Alford's study has been replicated by Perry (1981) over the longer time period of 1930-79 and using a somewhat different statistical technique. The results were totally supportive of the Lewis-Beck and Alford conclusion.

> The research indicates that strong safety laws reduce coal mine fatalities and that, if laws are strong, coal mine fatalities decrease with increases in federal spending on mine health and safety (Perry 1981, p. 13).

One of the problems with time-series analyses such as the foregoing is that we are limited to comparing fatality rates because, while the meaning of death is relatively invariant across time, the definition of nonfatal accidents has varied enormously in response to changes in workers' compensation law and the like. If we wish to investigate the impact of regulation on nonfatal accidents, we must do a cross-sectional analysis (or a pooled cross-section) within one of the relatively short periods of time when accident definitions remained constant. Such a study has been undertaken by Leslie Boden (1983) of Harvard in a pooled cross-sectional analysis of the 12 quarters during the 1973-5 period on 535 American coal mines. Boden found, after controlling for seam thickness, mining technique (continuous, conventional, longwall), mine size, unionization, and whether the firm was a captive (owned by a firm that used its output), that a 25 percent increase in inspections was associated with a 7 percent to 20 percent reduction in fatalities. Boden's model also estimated that a 25 percent increase in inspection days per section would decrease disabling injury rates per million tons of coal by 15 percent to 20 percent and total (disabling plus nondisabling) injury rates by 12 percent to 20 percent. These estimates do not, of course, include the impact inspectors have on the abatement of health problems (coal dust, noise, exposure to dangerous chemicals) as well as accidents. The Boden study— a sophisticated analysis on a large sample of observations— provides further powerful evidence that inspection substantially improves safety for miners.

Perhaps the most potent evidence of all for the efficacy of inspection comes from the "resident inspector program" introduced by MSHA in the mid-70s and abandoned at the end of the decade. The basic idea was to station a full-time MSHA inspector at those mines where hazards were seen as severe, or where the need for an inspector's daily presence was great, for example, at a mine so large that only a full-time inspector could ensure coverage of all the sections. Mines that released excessive methane were also likely to get a resident inspector. The guidelines also specified that mines would be included in the program if they were:

> mines that have been designated for frequent spot inspections under Section 103 (i) of the Act, have a lost day injury incidence

rate greater than the national average, and scored less than 80 percent on their mine profiles under the Mine Profile Rating System.

The Mine Profile Rating System was a method for scoring mines according to their hazard level and the adequacy of company safety policies, procedures, and resources for dealing with those hazards. In short, mines went on the program because they had a bad safety record. It is remarkable then that, once mines were given a resident inspector, their average accident rates fell well below the national average. In the first full year of the program's operation (1976) this was not true for fatalities— the seventy-six mines on the resident inspector program had a fatality rate 44 percent higer than the national average. This was a result of the Scotia mine (a mine with a resident inspector) blowing up, with the loss of 26 lives. Exluding Scotia, the mines with a resident inspector had a 50 percent lower fatality rate than the industry total. In 1977, it was 80 percent lower overall, and in 1978, 33 percent lower. For total injuries (degrees 2-5) the superior performance of mines on the resident inspector program was less dramatic. Perhaps one reason is that the presence of an inspector also encourages the reporting of accidents. In 1976, the seventy-six resident inspector mines nevertheless had a 9 percent lower injury incidence rate than the national average. In 1977 it was 10 percent lower, and in 1978 only 1 percent lower.

This program shows that you can take high-risk mines with a high accident rate and turn them into mines of below average risk simply by having an inspector reside full-time at the mine.

The evidence reviewed in this section in convincing that the tremendous improvement in mine fatality rates over the past 100 years were not simply a matter of technical progress or of workers inevitably demanding more safety as they became more affluent (Viscusi 1983). Safety improved during historical periods when safety enforcement was not being strengthened. The most dramatic periods and places of improvement, however, have been associated with the strengthening of government enforcement efforts.

The astute reader will have noticed that we have not so far demonstrated that violations cause accidents and inspections reduce violations, but rather that violations cause accidents and that inspections reduce accidents. We have skipped a stage in the inferential process. Unfortunately, there seem to be no direct studies of whether inspections reduce violations. There is good reason for this: the only way to measure violation rates is by inspection; therefore, the dependent variable (violation rate) is hopelessley confounded by the independent variable (inspection frequency and diligence).

## Does Punishment Prevent Violations?

We know: (a) that if we can prevent safety violations we can save lives, and (b) that what inspectors do saves lives. We do not know, however, that it is the punitive aspects of what they do (as opposed to their educative and persuasive endeavors) which protect miners; nor do we even know whether it is their effectiveness in preventing violations that is responsible for their success. There will probably never be satisfactory empirical evidence on these questions, because increased punishment and increased persuasion (which both happen when inspection increases) can never be separated out. We will see, for example that this is the bottom line of the sweeping historical analysis in the next paragraph, and it would also be the downfall of more finely tuned microstudies of the problem, if any social scientist were brave enough to attempt them.

Chapter 1 showed that the twentieth century has seen a British trend toward decreasing punitiveness in coal mine safety regulation, while the United States was becoming more punitive. Since both countries have shown substantial declines in fatality rates over the century (although Britain has always had lower rates than has the United States—in both its punitive and non-punitive eras, one is tempted to conclude that fatality rates move independently of the punitiveness of enforcement. But such a conclusion would be mistaken; in fact, in both countries the most rapid declines in fatality rates occurred during periods of increasing punitiveness. In Britain, the turning point was the 1910-20 decade. From the mid-nineteenth century to that decade, the use of prosecution to achieve compliance had been increasing; but after World War I, as we saw in chapter 1, prosecution was used less and less frequently. From figure 5 it is clear that the steepest descent in fatality rates occurred between the middle of the nineteenth century and the decade preceding World War I, the years of growing punitiveness, while after that the decline was more gentle. Equally, in the United States, the periods of most rapidly growing punitiveness, most notably after 1969, were the periods of greatest decline in fatality rates (see fig. 5 and Lewis-Beck and Alford [1980]).

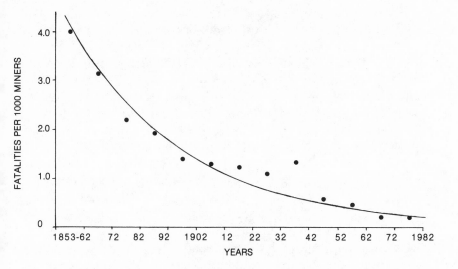

*Figure 5.* United Kingdom: Fatalities per 1000 miners employed from 1853.

Even though for both countries greater use of prosecution has been historically associated with an accelerated decline in fatality rates, does this really demonstrate that punishment reduces fatalities? The problem is that increased punitiveness never happens on its own. In the years following the 1969 act in the U.S. the use of fines and withdrawal orders to sanction mine operators grew substantially, but there was also considerable growth in the number of inspectors. One cannot rule out the possibility that the drop in fatalities was totally the result of the increase in the persuasive pressure put on operators by this swelling cadre of inspectors, rather than of the increase in punitive pressure. Because inspectors use a variable and indeterminite mix of punishment and persuasion, any effects flowing from a growth in inspections are not necessarily the result of punishment. Historically, increased punitiveness toward mine safety violations has generally reflected increased public concern over death in the mines, but this public concern tends simultaneously to engender greater efforts to persuade operators to run their mines safely, greater investment in training and safety technology, and, indeed, more responsible self-regulation.

In the end, we only have satisfying evidence on two of the three propositions required to justify the conclusion that punishment saves the life and limb of coal miners. There might be evidence consistent with the proposition that it is the punitive aspects of what inspectors do that saves lives, and we have certainly found no evidence to refute this proposition but any conceivable type of data in this area is open to multiple interpretations.

### Why Both Punishment and Persuasion Probably Work

Knowing that both inspection and preventing violations succeed in saving lives, it would be foolish to respond to our uncertainty over whether the punitive or the persuasive aspects of inspection produces this result by doing away with either punishment or persuasion. We might just do away with the more important one, or with the less important one without which, nevertheless, the more important cannot be viable. For example, persuasion might work best when everyone knows that inspectors can and do resort to punishment if their persuasive overtures are ignored.

Furthermore, the accumulated wisdom from the study of business regulation generally, beyond the limited domain of coal mine safety, instructs us that both punishment and persuasion are vital. It is to this accumulated wisdom that we now turn. First, we will consider arguments for the conclusion that punishment prevents corporate law violation.

### The Case for Punishment

It is relatively uncontroversial to argue that we ought to punish people who rob, rape, steal, or murder. Yet, punishment directed at these kinds of offenders is far less likely to prevent crime than punishment directed at corporations that violate the law. One reason that punishment directed against common criminals, and particularly young offenders, can backfire is the stigmatizing of the punished individual. Labeling theory contends that punishment pushes delinquents deeper and deeper into a criminal self-conception (e.g., Becker 1963). Proponents of this view cite as support the unexpected findings of the Cambridge longitudinal study of delinquency by West and Farrington (1977). They found that boys apprehended for, and convicted of, delinquent offenses became more delinquent than boys who were equally delinquent to begin with but who escaped apprehension. West and Farrington speculated that the reasons for this may be that labeling a youth as delinquent further alienates him from teachers and employers, and discourages more respectable companions from associating with him. Indeed, the whole paraphernalia of the criminal justice system can confirm a youngster's self-identification with delinquent groups.

Directing punishment against corporate offenders, or against individuals responsible for corporate violations, simply does not backfire in this way. Corporate offenders tend to regard themselves as unfairly maligned pillars of respectability, and no amount of stigmatization is apt to convince them otherwise. One does meet people, who as a result of labeling, have developed a mental image of themselves as a thief, a safecracker, a prostitute, a pimp, a drug runner, and even a hit man, but how often does one come across a person with a self-conception as a corporate criminal?

Chambliss (1967) argues that white-collar criminals are among the most deterrable offenders because they satisfy two conditions: they have no commitment to crime as a way of life, and their offenses are instrumental, rather than expressive. Corporate violations, such as those of safety laws, are almost never crimes of passion; they are neither spontaneous nor emotional, but lie more in the domain of calculated risks taken by rational people. As such, they should be more amenable to control by policies based on the utilitarian assumptions of the deterrence doctrine.

Individual corporate offenders are also more deterrable because they have more of the valued possessions one can lose through a criminal conviction, such as social status, respectability, money, a job, and a comfortable home and family life. As Geerken and Gove (1975, p. 509) hypothesize, "The effectiveness of [a] deterrence system will increase as the individual's investment in and rewards from the social system increase." Clinard and Meier, moreoever, place particular emphasis on the "future orientation" of white-collar offenders:

> Punishment may work best with those individuals who are "future oriented" and who are thus worried about the effect of punishment on their future plans and their social status rather than being concerned largely with the present and having little or no concern about their status. For this reason gang boys may be deterred by punishment less strongly than the white-collar professional person (1977, p. 248).

In general, the arguments about the deterrability of individuals convicted of corporate crime are equally applicable to the corporations themselves. Corporations are future oriented, concerned about their reputation, and quintessentially rational. Although most individuals do not possess the information to calculate rationally the probability of detection and punishment, corporations have information-gathering systems designed precisely for this purpose.

This rationality of corporations means that they are eminently deterrable, but they are also difficult to deter in practice unless penalties are severe enough to make it worth the corporation's while to comply. This is the perennial problem with fines as a sanction against corporations. For mammoth corporations, setting a fine large enough to have a deterrent impact is almost impossible.

> The $7 million fine which was levied against the Ford Motor Company for environmental violations was certainly more than a slap on the wrist, but it rather pales beside the estimated $250 million loss which the company sustained on the Edsel. Both represent environmental contingencies which managers are paid high salaries to handle. We know they handled the latter—the first seven years of the Mustang more than offset the Edsel losses. One can only infer that they worked out ways to handle the fine too (Gross 1978, p. 202).

The ineffectiveness of fines as sanctions against coal-mining companies in the United States is clear. In 1981 the average fine imposed was $173. Little wonder that a survey of mine operators found that:

> About 90 percent of the operators stated that civil penalties assessed or paid did not affect their production or safety activities. Penalty dollar amounts were not considered of sufficient magnitude to warrant avoidance of future penalties and improvement of safety procedures. Civil penalties were classified as a "cost of doing business," or as a "royalty" paid to the government to continue in business. Penalized producers saw no connection between penalties and safety. Only about 10 percent of the operators, all small, claimed to avoid violations because of fines (Mine Enforcement and Safety Administration 1977, p. A2).

From a sample of 76 underground coal mines over a period of two years, the same study failed to find any significant correlation between the amount of civil penalties imposed and improvement in injury rates (Mine Enforcement and Safety Administraton 1977, p. B3-5), a finding confirmed by Boden (1983). Hardly a surprising finding, given how inconsequential are even the most severe fines imposed. The penalties are also rendered ineffective by a failure of corporations to pay them in a large proportion of cases. At the time of the Buffalo Creek waste tip disaster, in which more than 125 people lost their lives, the operator had been assessed fines exceeding $1.5 million, not a cent of which had been paid. Prior to the Blacksville disaster, in which 9 men lost their lives in a fire triggered by a trolley wire ignition, the operator had been assessed for 379 violations, 178 of which were for electrical or trolley wire standards, and, of $76,330 assessed, only $31,090 had been paid (Subcommittee on Labor 1978, p. 43). Naturally, the legal costs to the government of going to court to compel collection of many of these petty fines are not worth the candle.

In the survey of the Mine Enforcement and Safety Administration discussed above, nonfinancial aspects of the civil penalties were mentioned as more important than the fines themselves.

Pride was a recurring issue. Operators perceived violations as
"black marks" on their safety record, so violations were avoided
not because of the associated penalty but for the sake of pride
(1977, p. A3).

This finding conforms with the results of seventeen case
studies of the impact of allegations and prosecutions against
large companies, by Fisse and Braithwaite (1983). The impacts
that were genuinely felt by the companies tended to be not so
much financial, even in cases where large fines were levied, as
the more intangible consequences of adverse publicity for
corporate prestige and employee morale, the harrowing experiences of senior executives in dealing with protracted cross-
examination, and the dislocation of top management from their
normal duties while they defended the corporation against
public attack.

But even these nonfinancial impacts are minimal or
nonexistent when it comes to the imposition of civil penalties
for coal mine safety violations. Civil penalties for such offenses
are so routine in the United States today, with 140,000 imposed a
year, that they attract no publicity whatsoever. This contrasts
with the situation during Britain's punitive era. The Royal
Commission on Safety in Coal Mines reported:

We recognise that the relative infrequency of prosecution
enhances the importance attached to legal proceedings when they
do occur. The prosecutions which are taken by the inspectors of
mines attract a great deal of local attention; they are fully reported
and discussed in the local Press; and a conviction carries with it a
correspondingly heavy stigma, which is not measured merely by
the amount of the penalty imposed (1938, p. 85).

We will return later to the theme that the frequency of
fining in the United States today has devalued the currency of
punishment. However, at this point it must be added that, on the
rare occasions when the U.S. government decides to prosecute a
safety violation through the criminal courts instead of through
their administrative civil assessments system, stigma sometimes
occurs as a result of reporting by the local press in the coal-
mining regions.

To summarize, there is no reason why coal mine safety violations, like other largely rational corporate economic behavior, should not be readily deterrable. Deterrable though they may be, they have not often been deterred in practice because of paltry fines, the ability to pass fines on to consumers in higher energy prices, the failure to collect fines, and the devaluing of stigma by making the imposition of civil penalties routine and automatic. This is not to say that even these automatic civil assessments could not be made stigmatic. One of the recommendations of the National Academy of Sciences report hints at one route to stigma:

> *Encourage publication of annual ranking of companies by their injury rates.* We have seen evidence that publication by the President's Commission on Coal of the rankings by injury rates of the 20 largest companies had a perhaps unintended influence— namely, managers of these companies were concerned as to their standing relative to other companies with respect to safety. Embarassment from being placed on the lower half of such a ranking could be an inducement to the managers of such companies to try to improve their standings (1982, p. 15).

Indeed, MSHA has at its fingertips the data with which to put out a press release each year, naming the three companies with the worst safety records and the worst violation records. Equally, each year they could bestow favorable publicity on the company with the best compliance record and the best safety record (in fact, they already do the latter in collaboration with the American Mining Congress).

I do not want to suggest for a moment that, because of the way punishments are imposed today upon the American coal-mining industry, deterrence is never achieved. Mention has already been made of the occasional use of criminal prosecution. More important, however, are some 5,000 withdrawal orders imposed annually by federal inspectors. These often have severe financial consequences for the operator, depending on the length of time a section is prevented from cutting coal. Once a section has been shut down as dangerous, it is amazing how quickly an operator who has been procrastinating for months will comply.

The withdrawal order is an instructive example; it shows that there are reasons other than deterrence to explain why corporate illegality is more preventable than common crime. Of course, the main reasons for withdrawal orders is not deterrence but the protection of miners who are being required to work in hazardous conditions. We can totally and easily incapacitate a criminal foreman who runs his section in flagrant disregard of the law simply by shutting down his section. In jurisdictions where foremen are required to have certificates of competence, we can also incapacitate him by taking away his certificate.

The resident inspector program is another example of an effective strategy for incapacitation. Flagrant violation of the law is more difficult when a government inspector is based full-time in the mine. The viability of incapacitation as a strategy for preventing mine safety violations contrasts starkly with the dismal failure of that doctrine for common crime. There are times and places when incapacitation for rapists has been attempted by castration and for pickpockets by cutting off their hands. Today we attempt to incapacitate by throwing common criminals in prison. A considerable volume of empirical research shows the enterprise to be a failure largely because we cannot predict who will be repeat offenders and hence who should be locked up and who should be kept out of our scarce and expensive prison spaces (Conrad and Dinitz, 1977; Van Dine, Conard, and Dinitz 1979; Cocozza and Steadman 1978; Cohen, Groth, and Siegel 1978).

In addition to deterrence and incapacitation as successful strategies in safety enforcement, rehabilitation—another strategy that has largely failed with common crime (Lipton, Martinson, and Wiles 1975)—is a workable doctrine for preventing corporate crime, including mine safety offenses. Many corporate violations arise from defective control systems; insufficient checks and balances within the system to ensure the law is complied with; poor communication; and inadequate standard operating procedures which fail to incorporate safeguards against reckless behavior. All this has been clearly illustrated by our analysis of the reasons for coal mining disasters, in chapter 2. A growing literature, particularly from the Australian school of empirical corporate crime research, shows that voluntary rehabilitation of defective standard operating procedures is a common response to prosecution of corporations (Hopkins 1978; Waldman 1978; Fisse and Braithwaite 1983; Braithwaite and Geis 1982). Sometimes corporations do this to get regulatory agencies off their backs, at other times simply from a genuine concern to put things right. If corporations do not voluntarily rehabilitate, they can be prodded or forced to in several ways: by consent decrees negotiated with regulatory agencies; probation orders placing the corporation under the supervision of an auditor, safety engineer, or other expert who would ensure that an order to restructure compliance systems was carried out; or by withholding sentencing of convicted corporations until they produced a report for the court on the weaknesses of their old compliance systems and agreed to implement new ones. The kind of rehabilitative measures that might be mandated include the appointment of a new vice president for safety; strengthening the powers of company inspectors to ensure that they cannot be overruled by unscrupulous production-oriented mine managers; institution of a corporate safety auditing program; increasing the powers of union-management safety committees; demoting incompetent managers from positions critical to safety performance; writing and enforcing new plans to counter a hazard; requiring investment in new safety technology; restructuring reporting relationships and the duty to know; modifying bonus payment schemes that give inordinate incentives to production but none to safety; and so on.

Rehabilitation is a strategy more workable with corporate crime than with common crime because criminogenic organizational structures are more malleable than are criminogenic human personalities. A new internal inspection group can be put in place much more readily than can a new superego. Moreover, state-imposed reorganization of the structure of a public company is not so unconscionable an encroachment on individual freedom as is state-imposed rearrangement of a psyche.

Since deterrence, incapacitation, and rehabilitation are all such viable doctrines in punishing mine safety violations, and since the American experience shows that obtaining convictions is not difficult, punishment must be given an important place in any approach to preventing offenses. Civil convictions on the basis of proof on the balance of probabilities has proved easy, and more surprisingly, MSHA has won convictions in all twenty-eight criminal cases that have gone to trial since 1978, a success rate unrivaled in the regulation of business crime.

As a result of this efficacy of deterrence, incapacitation, and rehabilitation, there is a much stronger case for reliance on punishment as a preventive mechanism with mine safety violations than there is with common crimes of theft or violence. How many criminologists would dare suggest that the dramatically increased resources we have devoted to enforcement of the law over the last fifty years has succeeded in reducing crimes against persons and property? Yet even the Bituminous Coal Operators Association in the U.S. has conceded that:

> The two most dramatic reductions [in coal mine fatalities between 1961 and 1976]—in fatalities caused by roof and rib falls and by gas and dust explosions—were undoubtedly due to the combined efforts of vigilant enforcement of the Act and hundreds of thousands of manhours spent by industry operating and safety personnel... (1977, p. 4).

## THE CASE FOR PERSUASION

### Cooptation or Pragmatism?

In coal mine safety enforcement, food and drug regulation, or pollution control, an inspector's initial commitment to punishment tends to be weakened in favor of preference for persuasion. In general, inspectors move toward a more sympathetic stance toward business. For example, Kelman (1981, p. 184) presented American Occupational Safety and Health Administration inspectors with two statements at the extremes of a scale: "Most employers are sincerely interested in assuring safe and healthful workplaces for their employees," and "Most employers are mainly out to make a buck, and will do only the minimum necessary to improve employee safety and health." After choosing one of these alternatives, inspectors were asked, "If someone had asked you the above questions *just before you started working for OSHA*, where would you have placed yourself on the seven-point scale then?" The results showed a dramatic shift toward greater sympathy for the regulated.

Why does experience in the field produce such shifts in inspectors' sympathies? Three interpretations seem possible:

1. Inspectors are coopted by business.

2. Simplistic stereotypes of employers as rapacious profit-maximizers break down when inspectors meet sensitive, concerned individuals in management roles, who do not conform to the stereotypes.

3. Inspectors learn in practice that when dealings with employers are conducted on the basis of assumptions that they are responsible and have a will to comply, greater compliance results than through dealings under assumptions of bad faith.

There is undoubtedly some truth in all three interpretations. Cooptation is a reality. Inspectors observe the careers of aggressively punitive colleagues falter under the pressure of constant complaints from industry to their supervisors for "unreasonableness." If these punitive inspectors get fed up with the restraints of agency superiors, where are they to turn for alternative employment? Since coal mining is all they know, only good will from mine operators can get them a job with earnings to match their expertise. This is especially true if one has a house and settled family in an isolated mining community where the mines are virtually the only employment alternative.

More blatant means can be used—witness the Appalachian operator who jokingly told an inspector, "For $500 I could get a New York hit man to take care of you." The inspector thought he was "only half joking." But the predominant pressures for cooptation are subtle. No one likes to be an ogre; no one enjoys spending working days surrounded by people who barely conceal their contempt for you. One inspector explained that, after inspections in which many violations were cited, managers have taken him aside and said, "What have you got against us personally?" Such interpersonal pressures tend to keep an inspector "reasonable," as that word is interpreted by the people the inspector must live with daily. In a small mining community, moreover, it is not only from nine to five that the inspector must live with those whom he regulates.

I got a real feeling for the great personal difficulties a good inspector experiences when I went into a mine with one American inspector. Our arrival at the mine was unexpected, but the mine's safety officer went to some trouble to secure a vehicle for us so we would not have to walk the tunnels during the inspection. When the vehicle arrived, the inspector immediately checked its brakes and a couple of other features and told the safety officer to get something minor fixed. His reply was to turn to the driver and say sarcastically, "Damn good of you, Fred, to come out of your way to give us a ride." Even union check inspectors have to put up with such jibes from their own members as "Here comes little Hitler" (Hopkins 1984, p. 30).

While careerist and interpersonal pressures do compromise punitiveness, this is not the total explanation for cooptation. Experience does teach inspectors that their stereotypes of the rapacity of mine managers are often wrong and that preconceptions of an irreconcilable conflict of interest between mine safety and profits are frequently misplaced. This happens, for example, when the inspector encounters such companies as Bethlehem Steel which, as we saw in chapter 3, have a pervasive company philosophy that improved performance is in the interests of profits and that a company with a strong and genuine commitment to safety, even though it might sometmes lose money because of it, will be ahead in the long run.

For two basic reasons it is often reasonable of an inspector to proceed on the presumption that mine operators have a will to comply with the law. First, there is the reality, alluded to above and in previous chapters, that the costs of accidents are enormous, many times the cost of insurance premiums (Grimaldi and Simonds 1975, pp. 391-422), and that, therefore, willingness to heed the inspector's advice can often save the company money. Even fires, explosions, or roof falls that fortuitously do not cause any injury can cost millions in damage to working areas. Second, it is probably true that fewer accidents are fundamentally explicable by callous profit-maximizing behavior than by careless indifference to dangers that do not seem very real to people at the time. Hopkins has captured well the psychology of miners' safety consciousness:

An explosion is a rare event. Most mines have never had one and several years may elapse without a single explosion anywhere in the Australian coal fields. Thus the miners' own experience leads them to discount the possibility of an explosion as they go about their work and to view the limits and standards specified in the legislation as somewhat arbitrary and unnecessarily stringent. They know from their own experience that gas can be tolerated with impugnity and they tend to see the standards laid down in legislation as having been devised by people with little or no knowledge of actual mining conditions. What happens, in other words, is that, based on their own experience, miners develop their *own* safety standards which diverge markedly from the official ones. Unfortunately, since the experience of any particular group of miners is limited, their own experience-based standards do not take into account the exceptional events or circumstances which official standards are designed to allow for (1984, p. 32).

Most mine operators would agree that, because of their own preoccupation with problems having a high probability of materializing, they need safety inspectors who will constantly remind them and their employees of consequences which, while perhaps relatively improbable, would be disastrous. In other words, competent mine managers realize that they and their workers do fall victim to standards based on their own limited experience and to policies which allow day-to-day problems that *must* be solved too much momentum, while safety catastrophes that *might* happen are neglected. Put another way, the rational manager views safety advisors as a necessary frustration in the same way that he approves of the aggravation of legal counsel who advises against precipitous actions with improbable consequences beyond the manager's practical experience.

The rational manager cannot hope to weigh up all the possible consequences of a given course of action. His attention span is limited to likely consequences routinely dealt with. Specialists, such as lawyers and safety officers, must be relied upon to tap him on the shoulder if there are, outside this frame of reference, unforeseen or forgotten contingencies that could produce serious results. Even in such everyday activities as breaking the speed limit to keep an appointment, we all need to be pulled up from time to time for the universal human failing of giving undue predominance to immediate goals. Rational managers have the good will to listen to what the safety inspector has to say because they know that, if fatalities are to be avoided, they must have these reminders.

The punitive-adversarial model succumbs to too rational a view of business decision making. It assumes that managers have perfect information about the economic benefits from breaking particular rules and about the probability of accidents from the violation. They are then assumed rationally to trade off the loss of life and limb against profits. In fact, in particular situations the manager often has neither type of information. Instead, consciousness of remote and poorly understood safety consequences is suppressed in the immediate struggle to get on with the job. Rational management, realizing that there is perhaps one chance in a thousand that such suppression will result in their mine being blown up, and realizing that a general policy of commitment to safety will save money in aggregate even if it does not in particular cases, will have a policy of heeding the government inspector, unless the circumstances are exceptional.

Of course, not all corporations are so rational, just as all citizens do not rationally accept that they benefit from the enforcement of traffic speed limits. For ideological reasons that in a microeconomic sense might be irrational, some operators resent government inspectors simply because they are an intrusion on unfettered free enterprise. They also foolishly adopt the view that production is the only thing that matters, and that any safety edict which sets back production is to be bitterly resisted.

Inspectors meet this kind of manger as well, and perhaps when they meet many such, they join the ranks of those who remain more punitive, whereas the inspectors who meet managers inculcated with a Bethlehem Steel philosophy are more attracted to persuasion.

We have therefore established a case that at least some of the time the inspector is confronted with managers who for quite rational reasons have the good will to comply. But this good will does not always arise from economic rationalty. It is also frequently engendered by a genuine concern for the safety of miners who are friends and whose families are known to the manager. Whatever the reason, talk to most inspectors, even union check inspectors who are members of the Communist Party of Australia, and they will tell you that many or most operators, though certainly not all, have the good will to heed advice and a desire to comply with most laws.

## Building Commitment to Comply

When there is willingness to do the right thing, across-the-board punishment is simply not the best strategy for maximizing compliance. Punishment is the best strategy when good will is wanting. We apply this common sense psychology in educating our children, in management, and in our everyday lives. For example, if we want to get our spouse to do the washing up, and we have a spouse who basically accepts the principle of doing a fair share of the housework, we find that retaliation is more likely than compliance if we say, "Unless you do the washing up I won't do the cooking tomorrow night." Most of us find that success is more likely if we appeal to our spouse's better nature. Punishment is something we resort to only when we confront a spouse, a student, or a colleague at work to whose better nature we cannot appeal for compliance with the goals we have in mind. If there is one thing that people who fail as spouses, teachers, or managers have in common, it is their inability to understand that you don't try to achieve goals by punishment until you've first tried appealing to peoples' better nature. Yet this is the very mistake that the U.S. Mine Safety and Health Act perpetrates. It imposes automatic penalties for any violation of a mandatory health and safety standard. There is no discretion for the inspector to give operators a second chance through a warning. Every schoolteacher knows that in some circumstances a child who would have been alienated by punishment can be given a greatly enhanced will to behave by saying, "That's not like you, Johnny Brown," and then forgiving the transgression. This strategy is no good with students who haven't a flicker of will to behave; with them a second chance will only be interpreted as weakness.

Unfortunately, many critics who favor a punitive-adversarial approach to business regulation are unwilling to make these distinctions. They want uncompromised and consistent punishment of corporate wrongdoers. The price they must pay for such indiscriminate use of punishment is lower compliance, because by remorselessly punishing those with a genuine desire to comply, we alienate them; by rejecting opportunities to give sincerely motivated managers a second chance, we forego the opportunity to build a commitment to try harder to ensure compliance in future.

At its worst, an uncompromising punitive strategy can lead to what Bardach and Kagan call an "organized culture of resistance"—a culture that facilitates the sharing of knowledge about the methods of legal resistance and counterattack. As an example, Bardach and Kagan cite the advice of one legal expert to appeal all OSHA citations, not just those to which companies object strongly, so that they can "settle a case by giving up on some items in exchange for dismissal by OSHA of others. Those who leave certain things uncontested are needlessly giving up this possibility" (1982, p. 114).

Punishment and persuasion are based on fundamentally different models of human behavior. Punishment presumes man to be a rational actor who weighs the benefits of noncompliance against the probability and costs of punishment. Persuasion presumes man to be reasonable, of good faith, and motivated to heed advice. Neither model fits very well the situations mine safety inspectors confront in the field. But the punitive model is a better fit in some situations, and the persuasive model, in others. Hence, any philosophy of regulation that limits inspectors to either model (as American regulation effectively limits them to punishment, and British regulation effectively, to persuasion) will hamstring the efficacy of inspection.

The problem with persuasion is that, based as it is on a model of man as basically good, it fails to recognize that there are some who are not, and thus will take advantage of being presumed to be so. The problem with the punitive model of man as essentially bad is that we dissipate the will of well-intentioned people to comply when we treat them as if they were ill-intentioned. We need inspectors who have the common sense to select the right model at the right time.

Dissipating the motivation of operators to strive for safer mines is a disastrous consequence because the punitive law enforcement alternative can never fill the gaps left by the failure of persuasion. With coal mining, as with all complex areas of business regulation, one can never write rules to protect people against all the unsafe practices that can occur. Since building consensus to write new rules is a difficult and time-consuming process, since rule writing does not keep up with rapidly changing mining technology, and since every mine poses unique safety problems, government regulations never cover the field. The British, who have achieved the safest mines in the world, make the point that, if their inspectors enforced strict compliance with the Mines and Quarries Act of 1954 and the regulations arising therefrom, they would enforce a far lower standard of safety practice than they in fact do. It is persuasion, heeded by responsible managers, which achieves the higher standards.

Achieving better than the minimum standards set down in law is imperative, but inspectors will not succeed if punishment has been used with so little finesse that they lose their capacity to persuade. Perhaps one reason that the United States has such a shocking coal mine fatality rate is that trust and respect between inspectors and managers has been lost by blunderbus punishment policies. As the chief executive of the Bituminous Coal Operators Association said when interviewd: "Lives are lost because of inspectors with the paper syndrome and companies with the 'How do we minimize the violations?' syndrome." While there is more than a grain of truth in this statement, it is also true that lives are lost because of the failure of punishment to punish in an effective and discriminating way.

## Limits of the Letter of the Law

Enforcement of specific regulations comes to grips only inadequately with the poor work habits that are the cause of most minor accidents or the poorly conceived management plans that are the cause of most disasters (see chapter 2). The problem of poor work habits can only be dealt with by the job safety analyses and training programs that we saw the safest companies adopting, in chapter 3. Certainly we can enact a law that each miner have at least a weekly safety contact from his foreman (as the state of West Virginia has done), but unless management has been persuaded of its value we can expect the safety contacts to be perfunctory ("Work safely, Fred!")

Inspectors should assist job safety analysis by adopting a diagnostic and catalytic role. Bardach and Kagan underlined this point by quoting the safety director of a large corporation on what he thought OSHA inspectors should do:

> OSHA inspectors have the right to talk to employees. They'll go up to a machine operator and ask if everything is OK. What they really mean is, "Is there a violation I can write up?" If the man points out a broken electrical cord or plug, the OSHA guy will just write it up and put it on the list of citations.
>
> What they should do is this: He should ask the employee, "How long has it been that way? Did you tell your foreman about it?" He should call over the foreman and ask why it was still that way. Maybe the foreman will say, "I've told him three times... you're supposed to go to Supply and get a new cord." Then why didn't he? Maybe his job is set up so he can't. Maybe the inspector will find out there's no procedure for checking cords, or that there is, but that the employees don't know it well (1982, pp. 148-9).

It is conceivable that nitpicking punitive enforcement of specific rules can even corrupt the integrity of a total safety plan for a mine. For example, forcing a nongassy mine to drive an extra tunnel to come into compliance with ventilation rules might so weaken the roof that the roof control plan is compromised. It is easy to see how a regulatory game of cat-and-mouse whereby companies defy the spirit of the rules by exploiting loopholes, and government writes more and more specific rules to cover those loopholes, can ultimately lead to a rule making by accretion that gives no coherence to the rules as a package. Instead of dealing with the underlying problem (how well the total safety plan hangs together), regulatory cat-and-mouse leads to a barren legalism fixated at the level of specific regulations. Under a regime of legal gamesmanship, the temptation for inspectors to concentrate on simple and visible hazards rather than on underlying problems is especially severe.

This is not to say that there is no place for punitive enforcement of specific regulations in a sensible regulatory scheme. There is an important place for it, but one must be wary of what Merton (1968, p. 194) called "ritualism" in which the means to an end (specific regulations) became all-important at the expense of the end itself (improved safety), and we destroy the garden as a system by giving one of its plants too much water.

## Granting Dignity to the Offender

The competent inspector does not use command and control to achieve compliance unless he has to. Following Dale Carnegie's famous advice, he will get managers to do things by convincing them that it was their own idea to do so. Indeed, some of the time punishment may even be imposed with such finesse that the manager almost believes that the punishment was his own idea. To illustrate, an Australian inspector tells the story of persuading a manager that something which was not really required by the law ought to be done to make the mine safer. After writing the manager's agreement to make the change in the "record book" kept for the purpose at the mine, the interaction finished with, "If I come back and find it not done, we agree that work will be stopped until it is done." The manager, genuinely persuaded that the change ought to be made, concurs. A few weeks later the inspector finds that, because of a strike and other production problems, the manager has not got around to making the change. The manager is unhappy when the inspector stops production, but he hardly thinks he has been unfairly treated.

This is the kind of thing meant earlier by the suggestion that punishment ought to be used selectively and with finesse to back up persuasion. As in the above example, punishment can be extended in such a way as to maintain dignity and mutual respect between enforcer and offender. We can punish without labeling people as irresponsible, untrustworthy outcasts, but instead by inviting the offender to accept the justice of the punishment. Punishment can be administered within the traditional framework of assumptions of disharmony and fundamentally irreconcilable interests, or it can be imposed within a framework of reconcilable, even mutually supportive interests. Griffiths calls the latter the "family model" of punishment:

> The child gets his punishment, as a matter of course, within a
> continuum of love, after his dinner and during his toilet training

and before his bed-time story and in the middle of general family play, and he is punished in his own unchanged capacity as a child with failings (like all other children) rather than as some kind of distinct and dangerous outsider (1970, p. 376).

Family life teaches us that punishment is possible while maintaining bonds of respect, but surely only possible if we grant the offender integrity by trying to persuade him first. When we reach the point where it is justifiable to use the full power of stigma and make the offender an outcast, we must accept that, in achieving the goals of symbolizing harm and general deterrence by so doing, we also abandon all hope of rehabilitation by persuasion, at least for the time being. Perhaps the rapport necessary for persuasion can only be re-established when the regulatory agency gives some equally public recognition that the company is reformed—the company is decertified as deviant (Meisenhelder 1982).

### Incompatibilities between Punishment and Persuasion?

There is another important sense in which the goals of punishment and persuasion can be incompatible. Information about where compliance systems are breaking down is vital to making persuasion work, or, indeed, to making any approach to regulation work. The threat of punishment can have a chilling effect on the information-gathering process. In recent years, MSHA, as a routine matter, has sent a "special investigator" to fatal accidents. The job of the special investigator is to collect evidence so that criminal charges can be laid against any culpable individuals or corporations. Usually these special investigators are experienced miners who have undergone training in criminal investigation by police experts. At a serious accident they typically gather evidence by questioning individual mine personnel after giving them a form of Miranda warning. The interviews are usually taped. Some companies have responded by advising employees not to talk to the special investigator except in the presence of a lawyer from the general counsel's office.

Another reaction to MSHA's criminal program of the late 1970s has been a widespread refusal by companies, on Fifth

Amendment grounds, to fill out the section of MSHA's accident report form that asks for information about who was responsible for the accident.

The contention of the American industry is that, before the advent of "special investigation" procedures, accident investigations were generally frank and open discussions among management, miners, and enforcement officials. Such open exchanges, they condend, better achieved the main purpose of accident investigations—understanding what went wrong so that future occurrences of similar accidents could be prevented. Even though most accidents are carbon copies of mistakes that have caused accidents many times before, learning from mistakes is a more important justification for investigations than allocating blame. Undoubtedly the reading of Miranda warnings discourages some people from providing all the information in their possession. Nevertheless, at the end of the day, are "open and frank" investigations a more effective means of getting to the truth than inquisitions?

Rarely are people voluntarily so open and frank as to admit that incompetence or offenses of their own or of their friends were responsible for loss of life. They will talk about the negligence of their enemies, or causes of an accident that do not reflect badly on friends, but they will also do this under Miranda warnings. Mine safety law in Britain and New South Wales requires employees to answer any question put to them, while forbidding the use of their responses as evidence in any prosecution against them (except in a prosecution for making false statements). Similarly, in New South Wales, testimony given in judicial inquiries into coal mine fatalities cannot be used against those who give it. Yet, in the most recent three such judicial reports, the judge found that key witnesses had lied or covered up to protect their own reputations or those of others.

A counter argument can be mounted that when there is a real risk of criminal convictions the truth is more likely to come out. Would we, for example, ever have learned about the conspiracy in the United States to fix prices of heavy electrical equipment if certain individuals who were afraid that they themselves might end up in jail had not spilled the beans in return for a promise that they would not be prosecuted (Ermann and Lundman 1982, p. 91)? Had this investigation taken place as a nonpunitive, open and frank discussion, what incentive would there have been for these individuals to tell tales on their colleagues? For most organizational crimes, the only witnesses are members of the organization—there are no impartial outside observers who can give disinterested accounts of what happened. In practical terms, often the only way investigators can get to the bottom of what happened is by holding out the threat of prosecution to an individual and then withdrawing it when he testifies against others.

The Lockheed scandal is another example of how an inquisition can produce more revealing findings than an "open and frank" discussion. How did it all begin for Lockheed? It began when Northrop Chairman Tom Jones succeeded in relieving the pressure on his own company, before Senator Frank Church's Subcommittee of Multinational Corporations, by saying that Northrop's practices of using consultants to pass bribes were modeled on those of Lockheed. As the press turned their attention to Lockheed in response to the Jones bombshell, Lockheed's auditor's, Arthur Young, decided it was time for self-preservation. They asked top Lockheed executives to sign a letter stating in effect that, to the best of their knowledge, no bribes had been paid. Lockheed's executives were not prepared to let Arthur Young shrug off all responsibility by passing the buck to them. They refused to sign. Arthur Young, in turn, refused to certify the annual report. This brought on an inevitable government investigation.

In my research on corporate crime in the pharmaceutical industry (Braithwaite 1984, p. 473), informants told me of similar "protect your own ass" strategies. A quality control manager who fears that he may be held liable for an impure batch of drugs may "protect his ass" by writing a memo to indicate his personal opposition to the sale of the batch. The intention is that government investigators will find this opinion in the files, evidence which, while absolving the quality control manager, will help hang the company.

The safety director of one American coal-mining company conceded that he, too, occasionally engaged in this kind of "ass protecting": "I sometimes have to say to them [mine managers], 'You do what you please, but I'm telling you it's illegal.' Then

I'm covered." These four examples all show that fear of punishment flushes out information from actors who wish to protect themselves by telling tales on others.

Unfortunately, the aftermath of a fatality, for which no one wants to be blamed, even informally, is one of those occasions when we are unlikely to get to the truth by presuming good faith under the persuasion model—exploiting bad faith through threats and withdrawal of punishment is more likely to turn up the facts. Too often "free and frank"investigations are cosy cover-ups that coopt government regulators who have failed to enforce the law before the fatality. Conspiracies of disinformation are harder to hold together when fear of punishment can cause one conspirator to break ranks and tell the truth.

Another problem, however, is that a punitive regulatory framework can cause suppression of internal communication of accident information within the company and thereby inhibit self-regulation. For example, U.S. Steel's response to MSHA's criminal program has been to introduce a policy of destroying all internal fatality investigation reports not required to be kept by law. However, it appears that most companies do not have such a systematic policy of shredding reports. Some conceded to becoming more circumspect in what they write and, perhaps, to handling more of the communication involving allocation of blame by word of mouth.

At the other extreme, however, are companies that see the criminal program as an incentive to thorough documentation of vigorous internal investigation and rectification of problems. Those in the industry who hold this view argue that, once a company safety officer has written an adverse report on the failure of a mine to comply with the law, the fear that this report could be subpoenaed in a criminal investigation is a strong incentive to ensure that the violations are quickly rectified; the safety officer can then attach a follow-up memo to this effect. Again, here we have fear of punishment causing "ass protecting" behavior that enhances the internal communication of hazards and gives incentives to rectify them.

This second view of the appropriate response to the threat of company documents being subpoenaed was advocated by Messers Thompson and Sliter, of Hamel, McCabe and Saunders, in a paper, "Planning for Federal Safety and Health Inspections," prepared for the American Mining Congress.

> Some operators feel that more records mean more opportunities for MSHA to issue citations. While this possibility must always be considered in preparing documents, the advantages of maintaining and managing a thorough record-keeping system are compelling....
>
> Adequate record keeping can provide valuable evidence in defending certain enforcement actions or discrimination cases. For example, where an employee has complained about maintenance on a piece of equipment and requests an inspection by MSHA, complete records indicating the results of the company inspections and maintenance performed can provide the operator with the evidence needed to fight a citation or to support disciplinary action against a disgruntled employee....
>
> In summary, a thorough record-keeping system will help the operator to improve health and safety conditions in a mine, and it can provide a first line of defense against unreasonable enforcement actions (Thompson and Sliter 1981).

What is of some concern, however, is that fear of punishment may result in the structuring of communication blockages into organizations to ensure that the chief executive is not tainted with the knowledge of illegality. One vice president responsible for safety in an American coal-mining company was candid enough to assert his view that "the far left want to see company presidents put behind bars. So we have to protect them." This executive went on to explain that he learned that this was expected of him when he told the president he would send him a memo on a safety problem. The president replied, according to his recollection: "Don't send me a memo. Tell me if you have to. I don't want to learn what you do about it, but you fix it." So now the vice president responsible for safety realizes that "part of my job is to protect him" and that "the least amount of incriminating information must go to the president." If this is really a consequence of fear of criminal liability for the president, then it is very unfortunate indeed because safety is only likely to become a top priority in a company in which commitment to safety comes from the top down. Yet there is no way of knowing whether this kind of relationship between the vice president for safety and the president is typical or unusual in American companies. Curiously enough, in this company the president actively monitors the *statistical* safety performance of his mines so that managers of mines whose accident rates rise find themselves rapped over the knuckles by the chief executive.

In short, punishment can act both to block and to flush out knowledge and understanding of what causes safety breakdowns. Flushing out truth is a more common consequence of corporate punishment than suppressing it. In addition, a sophisticated punitive strategy can be designed to minimize suppression and maximize openness (e.g., incorporating protection rights for whistle blowers, exempting certain types of internal reports from routine government inspection [Braithwaite 1982b, pp. 1487-8], statutory duties to demand that instructions to compromise safety be put in writing—see page 159 below).

## Persuasion as Efficient Use of Resources

An important advantage of persuasion over punishment is that it is often a more efficient use of scarce regulatory resources. Is it really sensible for an MSHA inspector to follow the letter of the law and issue a citation when he comes across a piece of timber lying in the middle of a roadway waiting for someone to trip over it? All that paperwork and lost rapport from perceived pettiness for the sake of a $20 fine? It is far more efficient for the inspector to say, "Hey, get that timber out of the middle of the roadway!" or simply remove it himself, than to stop and write a citation (as the law says he must). By stopping to complete the paperwork, he gives himself less time to cover more of the mine and find violations of a more serious nature. A fundamental reason why British inspectorates moved away from a prosecutorial stance so early in the century was that they had devoted enormous resources to mounting prosecutions, only to find that courts imposed token penalties on the offenders (Rhodes 1981, p. 177). Until the attitude of the judiciary toward punishing corporate offenders changes, a policy of consistent prosecution of health and safety offenses makes limited sense.

Other American regulatory agencies have moved in a direction opposite to MSHA's by being increasingly selective in their use of prosecution on the grounds of efficiency. The FDA

(Food and Drug Administration) prosecutes far less frequently today than it has in decades past, mainly for reasons of efficiency (Heaviside 1980, p. 82). The change is regrettable, but understandable. Even regulators who start out with an almost missionary zeal for prosecution often conclude after some experience that they can better serve the public interest by opting for informal social control in a larger proportion of cases. The classic account is Phillip Schrag's (1971) description of his own metamorphosis after he took over the Law Enforcement Division of the New York City Department of Consumer Affairs. In response to a variety of frustrations, especially the use of delaying tactics by company lawyers, Schrag soon began to jettison his prosecutorial stance in favor of a "direct action" model. Nonlitigious methods of achieving restitution, deterrence, and incapacitation were increasingly resorted to. In addition to persuasion, these included threats and use of adverse publicity, revocation of license, writing directly to consumers to warn them of company practices, and exerting pressure on reputable financial institutions and suppliers to withdraw support for the targeted company. Much has been written elsewhere about the capacity of company lawyers to exploit the complexity of the law and assert every legal right of their clients, to such a point that prosecution becomes very expensive and time-consuming (e.g, Sutton and Wild 1978; Braithwaite 1980). This reality has led no less adversarial an official than the Securities and Exchange Commission's former director of enforcement, Stanley Sporkin, to conclude: "The trial of a case is like a war. You only have a war if you need it and can't solve it peacefully." (Wall Street Journal, 14 January 1981, p. 14).

To deal with the 140,000 citations its inspectors write in a year, MSHA has a system for assessing civil penalties without the necessity of going to court. Within the agency there is an Assessment Office, which uses a number of criteria to determine the size of the fine to be imposed for each cited violation, though since 1982 most of the de facto decision making about penalties has been going on in MSHA's district offices. Appeals against assessed penalties are common when they are relatively heavy, and a hearing before a judge is from time to time the ultimate result. Moreover, since there is widespread delinquency on the part of coal-mining companies in failure to pay fines, further court hearings to compel payment are also regularly required. The net result is an enormous drain on the resources of the agency for the imposition of what are puny fines. MSHA inspectors lose a great deal of time that could be spent inspecting mines as a result of the mountain of paperwork created by assessments:

> The inspector is required to prepare an assessment work sheet for every violation that he writes, and he is generally required to confer with the Assessment Office on one or more occasions concerning the facts surrounding each such violation. Once the case goes into administrative and/or court litigation, the inspector will inevitably be a witness since the government has the burden of proving the violation and the inspector is usually the only witness competent to testify concerning the violation. Because the inspector is inevitably a witness, he is frequently called upon in connection with pre-trial activities to answer interrogatories submitted by the operator or to give depositions. As the time for hearing approaches, he must prepare his testimony and he is generally required to be present with the MESA lawyer throughout the hearing (Bituminous Coal Operators' Association 1977, p. 35).

In addition to distracting inspectors from the task of inspection, the civil penalties program keeps 13 MSHA attorneys fully occupied, employs well over 100 people in MSHA's Assessment Office, which exists solely for the assessment of civil penalties (though this staffing level is declining), and keeps 13 administrative law judges and their staffs occuped with nought else but administrative hearings involving civil penalties. The levels of appeal against these paltry penalties are extraordinary. After the inspector issues a citation, company management has a first opportunity to challenge it at a mine close-out conference after the inspection. If this provides no joy, a health and safety conference at the district or subdistrict MSHA office can be requested. From there, the citation goes to the Assessment Office at MSHA headquarters for an independent determination of the penalty. The company can then appeal the Assessment Office decision to an administrative law judge, the administrative law judge  decision to the Mine Safety and Health Review Commission, the Mine Safety and Health Review Commission decision to a court of appeal, and up to the Supreme Court. One might well ask, If all this money were spent on employing more inspectors to check compliance in mines, would more lives be saved?

The cynics contend that operators use the appeal processes, particularly the new mine close-out and health and safety conferences, to keep inspectors out of the mines. On the other hand, these conferences can be seen as a remedy for the problem of punitive regulation that fails to help offenders to learn from their mistakes. This problem was highlighted by the interviews with miners and managers conducted by DeMichiei and his colleagues:

> Too often the MSHA inspector is viewed simply as a "police-man" whose only function is to identify violations and issue citations. The interaction between mine employees and MSHA inspectors generally consisted of being notified that an unsafe condition existed and corrective action was necessary. Very seldom were miners provided the opportunity to identify the root causes for the existence of unsafe conditions or the means to prevent their recurrence (DeMichiei et al. 1982, p. 32).

### Is Persuasion Morally Acceptable?

There are troubling moral concerns about persuasion as an enforcement strategy. Persuasion really amounts to negoti-ation—give and take. But can it be right for the law to be negotiable? Negotiation never works as a strategy to change behavior if one side does all the taking and no giving. As Nierenberg points out in his influential book on negotiation:

> Sometimes, when an opponent seems "on the run," there is a temptation to push him as hard as possible. But that one extra push may be the one that breaks the camel's back. Simply stated, one of the first lessons the negotiator must learn is *when to stop.... All parties to a negotiation should come out with some needs satisfied* (1981, pp. 29-30).

But should a safety inspector refrain from citing further violations, or fail to demand rectification of further deficiencies that endanger the lives of miners, simply because he believes he has already pushed management too far? It is acceptable to use what Nierenberg calls the "crossroads" strategy wherein you "introduce several matters into the discussion so that you can make concessions on one or gain on the other" (p. 117), when the matters being introduced into the discussion are violations of law?

It can be acceptable for the same reasons that we find negotiated justice in the form of plea bargaining and grants of immunity tolerable in law enforcement by the police. Although we regret that such compromises are necessary, we admit that by foregoing the opportunity to punish some crimes the police can ultimately use their scarce resources to provide better protection for the public. Equally, however, we judge that the horse trading must have limits. While we tolerate letting off a first-time juvenile shoplifter with a stern warning because the policeman has more important things to do than take the matter to court; while we condone a grant of immunity to a minor functionary of a criminal organization to entice him to testify against a Mr. Big; we surely should not countenance letting a known hit man go unpunished under any circumstances. Defining what sort of horse trading is acceptable in the enforcement of common crime is difficult and has never been done satisfactorily. However, defining reasonably clearly what kinds of violations of mine safety laws ought to be nonnegotiable can and has been done.

The U.S. Mine Safety and Health Act of 1977 defines certain types of violations as "significant and substantial" and other conditions as "imminent dangers." Until the Federal Mine Health and Safety Review Commission's decision, in Secretary of Labor, MSHA v. National Gypsum Company, there was a clear, meaningful, and appropriate definition of a "significant and substantial" violation. It essentially included all violations except those posing no risk of injury at all (purely technical or bookkeeping violations), or those violations involving only speculative or remote risks of injury. Since that 1981 decision, however, a violation is only significant and substantial if it can be shown that there is a "reasonable likelihood" that the hazard would result in an injury or an illness of a "reasonably serious nature." A broad category, tightly defined by specific exclusions, has been replaced by a vague narrow standard, ambiguous in its characterization of what is included. Whereas under the previous standard more than 60 percent of citations were significant and substantial, under the new standard (combined with Reaganite administration of the act) the proportion of citations which are significant and substantial has fluctuated between 22 and 42 percent.

The significant and substantial concept is sound. It is simply its definition that has gone awry in the United States. Citation of imminent dangers and knowingly committed, significant and substantial violations (under the old definition) ought to be nondiscretionary. Beyond that, citation of other violations, even significant and substantial ones that were not committed knowingly, could then be bargaining chips in negotiation.

We are still justified  in feeling uncomfortable about the latter, since a large disaster can arise from a confluence of many minor deficiencies in a safety system. But if the inspector can better ensure that the critical features of an overall safety system are in order by refusing to be nitpicking, or by giving the operator time to rectify certain matters that pose only low-level risks, then miners will be the beneficiaries of the uneasy compromise. There is no virtue in a self-righteous application of the law if the results are an inspectorate that spends more time in court than in mines, and an industry, alienated from a commitment to safety, resisting any improvement not achieved by legal compulsion.

## Inevitability of Reliance on Persuasion

There is a certain historical inevitability about shifts from formal punishment to negotiation in business regulation. Negotiation is not the normal way for a sovereign state to control private units. Organizations that are much more powerful than their adversaries seldom need to resort to negotiation. But as the adversary gains power more nearly equal to that of the controller, control through negotiation becomes the increasingly preferred option. We have seen this sociological reality in relations between employers and unions. There was a time when employers controlled unions through mechanisms of law. As recently as 1936, the Supreme Court of the United States determined that a minimum wage act was an unconstitutional interference with freedom of property (Friedman 1972, p. 55). Trade union activities were attacked on the basis of the law of conspiracy. But as trade unions became more organized and powerful, employers found negotiation rather than prosecution to be the only way to resolve disputes.

In the same fashion, governments find it increasingly difficult and costly to use the courts to impose sanctions severe enough to sting giant corporations. When government agencies, be they the Internal Revenue Service or the Environmental Protection Agency, attempt to change the behavior of organizations with the political-economic power of a U.S. Steel or DuPont, they realize that it is difficult for the state to be sovereign in its enforcement of the law. Negotiation between equals increasingly replaces law enforcement.

MSHA is unusual in the way it has swum against this historical tide by becoming more punitive. Other agencies deserve criticism for drifting too meekly with the tide, failing to find creative ways of imposing punishment that take account of the legal sophistication and burgeoning power of their adversaries. In contrast, MSHA can be criticized for drowning from exhaustion as it thrashes against the current, instead of staying afloat and making more deft use of negotiation to steer the direction as the coal industry is swept by historical and technological currents.

## Situational and Cultural Relativism in Enforcement

The relative merits of punishment and persuasion depend very much on the frequency of contact between inspector and management. Persuasion is a more potent strategy for coal mine inspectors than for factory inspectors, because the former have an ongoing relationship with management while the latter do not. Mine inspectors can nurture the rapport necessary for persuasion, because they typically visit the same mine at least a few times a year. They regularly return to see if their suggestions are being followed. In contrast, only a tiny fraction of factories are inspected yearly by a health and safety inspector, and if they are visited, it is usually by an inspector different from the previous one. In the U.S., while MSHA inspectors cover 3,000 mines, OSHA inspectors are responsible for 3 million workplaces. Because the factory inspector is such a ship passing in the night, he is unlikely to command much respect from management, and hence his power to persuade is limited. Deterrence through punishment therefore assumes greater importance as the way to change behavior.

The relative merits of punishment and persuasion also depend on the culture. Perhaps punishment is a more appropriate means of regulating business in the United States than in a society like Japan, or even Britain and Australia, because American business people give less deference to advice from government officials. The United States is unusual in the public hostility and mutual mistrust that obtains in business-government relationships. In most societies a powerful state bureaucracy preceded the flowering of industrial capitalism, and business leaders had to develop a close working relationship with their governments from the outset. The American state, in contrast, played no central role in guiding the industrial development of the frontier society. American business did not learn to cooperate with government, and when bureaucrats began to stick their noses into American business, a pervasive ideology of resentment toward government intervention took hold. The frontier society also produced self-assertive values— rugged individualism, which undermined willingness to defer to government. An adversarial relationship between business and government came to be viewed as healthy by both business and radical critics of business. Procedures for consulting industry about new regulations were modeled on adversarial trials rather than on small group discussion and consensus building.

Perhaps, then, American inspectors have a far more limited capacity to achieve compliance by persuasion than their colleagues in other countries. If a Japanese inspector tells a company that it must do something, it almost invariably does it. There is no question of challenging the government's right to issue the instruction or of litigating the legality of the order in a court of law. Sweden is a society with a more radical, critical attitude toward capitalism among its citizenry than the United States. Yet paradoxically, Sweden evinces an attitude of greater trust that companies will comply with the law without punishment and litigation. Kelman (1981, pp. 196-7) compared the attitudes about the untrustworthiness of business among

American and Swedish occupational safety and health in-spectors. They were asked to indicate where in a seven-point scale their attitude lay, when one end of the scale was defined by the statement, "Most employers are law abiding, and try to follow the standards simply because a government agency has issued them," and the other end defined by, "Without the penalty-imposing powers we have, many employers would simply ignore the standards." Fifty-six percent of the American inspectors, but only 15 percent of the Swedish inspectors, place themselves at the end of the scale defined by the latter statements. Thus, it may be that, while the punitiveness of American enforcement is responsible for the unwillingness of industry to defer to the advice of its government, the absence of deference to government may also explain the punitiveness. Each side is responsible for the predicament of the other.

## Conclusion

If the simple conclusion of the first half of this chapter was that punishment works in coal mine safety enforcement, the conclusions of this half is that persuasion works, although less so in some cultures than in others. Punishment can never operate on its own as an effective means of business regulation, primarily because so much of the harm that can be done in as complex an operation as a modern coal mine cannot be covered by up-to-date rules written by government.

The trick of successful regulation, then, becomes that of imposing punishment when needed, without undermining the capacity of inspectors to persuade. It has been argued that punishment and persuasion need not necessarily be incompati-ble so long as punishment can be imposed with a certain amount of discretion. Punishment can be imposed without labeling the offender as an outcast, without creating incentives to block open relationships of respect and trust between inspector and opera-tor, and, by highly skilled inspectors, even with such finesse that the operator accepts the punishment as justified. Because punishment depletes so many regulatory resources, strategies must be found to reserve punishment for the situations in which it will be maximally effective.

It is a mistake to believe that, under a more punitive regulatory regime, inspectors will be excluded from information on safety problems. When it comes to the crunch, the threat of punishment can be a more powerful tool for extracting truthful information about wrongdoing than persuasion. Thus, there is little reason for pessimism that information blockages will keep persuasion from working when it is mixed in with punishment.

The power to punish helps give legitimacy to regulators who wish to persuade. One is inclined to listen to the persuasive overtones of an inspector if the consequences of not listening is his replacing the velvet glove with the iron fist. The conclusion is therefore the same as that for the first half of this chapter: the complementarities between punishment and persuasion are more profound than the incompatibilities. The ensuing chapter suggests a strategy for a happy coexistence between punishment and persuasion.

# 5. When to Punish

In the preceding chapter it was argued that both punishment and persuasion are effective and necessary in regulating mine safety, that a punitive strategy can inhibit persuasion, and vice versa, but that this need not necessarily happen. Indeed, there can be more synergy than incompatibility between the two strategies. The task of this chapter and chapter six is to suggest approaches for extracting the most effective enforcement from a mix of punishment and persuasion.

Diver has captured the dilemma that brings us to this point:

> The marginal cost of deterring a particular harmful violation may exceed the harm that would be thereby avoided. One may reasonably assume that the marginal cost of deterring incremental violations increases, and the marginal benefits decrease, with the number of violations deterred. That is, the rational regulator will concentrate his efforts initially on violations that involve substantial risk of harm but are reasonably easy to detect and prove. As the scale of enforcement activity expands, the agency will be forced to seek out violations less readily observed or less harmful. At some point, the rising marginal cost curve and the falling marginal benefit curve presumably intersect. Beyond that point, further deterrence is counterproductive (1980, p. 264).

Fundamentally, further deterrence can be counterproductive in coal mine safety enforcement because litigation chews up resources that could be used on increased inspection and persuasion. In the United States this tradeoff is painfully clear; MSHA still does not achieve the four inspections per year for every coal mine in the country mandated by the Mine Safety and

Health Act. The situation does, however, appear to have improved dramatically from that of a decade ago when the General Accounting Office found only 31 percent of required safety inspections to have been completed in a sample of districts (General Accounting Office 1971, p. 83).

The solution proposed here to the problem of finite enforcement resources is in defining a hierarchy of regulatory response. At the bottom of the hierarchy, self-regulation should be granted in as many areas as experience proves the privilege not to be abused. When industry fails to respect the responsibility to regulate itself, then regulatory response can be escalated through a number of increasingly punitive and increasingly governmentally controlled strategies. These, in turn, will be described below as enforced self-regulation, command regulation with discretion to punish, and command regulation with nondiscretionary punishment. In addition to a hierarchy of regulatory strategies, a hierarchy of punitive response will be defined to ensure that the state has the leverage to reduce litigation by plea bargaining and/or by opting for less severe sanctions, which can be imposed without a full panoply of due process protections.

## SELF-REGULATION AND DEREGULATION

Regardless of how cynical one might be about trusting companies to regulate themselves, the hard reality is that, under even the most punitive and resource-rich regulatory scheme imaginable, most of the responsibility for making the system work will be left in the hands of the regulatees. Consider sampling of dust in mine environments to protect workers from black lung. MSHA currently collects 50,000 to 60,000 dust samples each year, while the industry collects 500,000 to 600,000 (General Accounting Office 1981, p. 53). Trusting industry to collect the samples might be putting the fox in charge of the chicken coop, but if the federal bureaucracy were to take over these half a million samplings per annum, the resources would have to be taken from somewhere else. Inevitably the result would be that fewer samples would be collected. This is perhaps why industry has sometimes supported suggestions that such costly monitoring functions as noise surveys be conducted by government officials rather than company employees (e.g., Consolidation Coal Co. 1980, p. 104).

Its superior resources is not the only reason that self-regulation can sometimes be a more potent alternative than government command and control. Other reasons are the superior investigative capabilities of companies and the more potent routine sanctions available to them. In the United States, MSHA and state inspectors are on average less competent and less well qualified than are company personnel with safety responsibilities. The reason is simply that the industry pays more than the government. One leading American company told me stories of three employees it had dismissed for either hazardous practices or failing to supervise safety responsibilities adequately; two of them became MSHA inspectors and the other, the head of coal mine safety enforcement with one of the states. The competence gap is not nearly so wide in Great Britain and Australia; there, inspectors are better qualified and better paid than many of their counterparts in industry.

Nevertheless, the superior capacity of company safety personnel to diagnose problems has less to do with their competence than their access to information. Generally, though not always, company safety personnel can expect greater openness from company employees than can government inspectors. This is because production employees know that top management wants them, and tells them, to be totally candid with company safety personnel, but admonishes them to caution in what they divulge to government officials. Often, though not always, the company safety person will spend more time and be more familiar with the peculiarities of a particular mine than will any government safety person who spreads his time across many mines. This combination of an average greater competence, access to information, support from top management, and familiarity with the mine means that it is the company safety people who have the superior capacity to discover who is responsible for wrongdoing.

Having discovered who was responsible, the company's safety department, perversely, has a greater capacity to punish them than has a government regulatory agency. In practical terms, individuals almost never go to jail for mine safety violations. There are only four examples of imprisonment in the

United States since World War II, and none in Australia or Britain. So, fines of up to $10,000 are the most potent routine sanctions against individuals available to the government. But a corporate safety department can do far worse to an individual than a $10,000 fine. It can recommend that a person be demoted or held back from promotion. This can cost a person many thousands of dollars for every year until retirement. The British are particularly strong on excusing their failure to prosecute violations in the courts by pointing out that a more potent punishment is to secure an assurance from the National Coal Board that a culpable manager will be transferred to "executive officer responsible for paperwork."

Private enforcement is not only potentially more potent because companies have investigative and punitive capabilities that are superior to those of governments, but also because private justice systems can more easily prove guilt. An employee charged under internal corporate discipline with failing to meet his safety responsibilities enjoys few due process protections. He can be convicted on circumstantial evidence that falls short of proving guilt "beyong reasonable doubt," and he may find that the burden is on him to prove his innocence. Should he choose to exercise a right to silence, he will find that the ultimate sanction of dismissal will be imposed on him anyway for so doing. I do not seek to laud private justice systems, which have such a history of abuse and scapegoating. Employees should be guaranteed certain of the rights under private justice systems that they enjoy under public justice systems. However, irrespective of reform to protect employees from unjust dismissal or demotion, proving culpability to the satisfaction of an employer will always be easier than proving it to a judge and jury.

All this means that it is a mistake to assume that self-regulation is *necessarily* a softer option than public enforcement. This observation has wider application. For example, in Australia it can be argued that doctors and lawyers who perpetrate fraud against medical benefits or trust accounts have less to fear from punishments handed down by courts of law than they do from being struck off by their profession or disqualified from government-subsidized business. Self-regulation will only be a tough and meaningful option where government demands guarantees that the price of greater freedom from government interference will be that the potent sanctioning capabilities of private justice systems will be unleashed.

It is proposed, then, that if a new area of regulation is to be introduced, unless there are persuasive arguments to the contrary, voluntary self-regulation should be the first regulatory strategy to be tried. If self-regulation fails, then the more interventionist strategies discussed below should be put into effect. For example, if the government has in mind requiring companies to undertake job safety analyses for all positions in coal mines, the first gambit should be to suggest that, unless the industry voluntarily introduces job safety analyses across the board, regulations will be written and enforced to compel such analyses. The agency would then wait for evidence of voluntary reform, of the punishment by corporate headquarters of managers who fail to implement the reform, and of trade associations urging the reform upon their members and withdrawing membership privileges from those who do not conform. If these reforms do not occur at a satisfactory level, the industry is told that it has blown its chance to self-regulate, and the rule writing begins.

This is not to say that self-regulation would be jettisoned in the face of evidence of some noncompliance, because significant noncompliance is normal under government command and control regulation as well. It is, rather, a matter of assessing whether the level and seriousness of noncompliance is sufficient to warrant a regulatory regime with more formal assurances. Within limits, deviance can be dealt with under systems of informal social control; mines that do not reform voluntarily can be blacklisted for blitz inspections since they have acted irresponsibly when given the chance to regulate themselves. Voluntary compliance therefore becomes the easiest way for the mine to get the inspectors off its back. This may smack of a bullying tactic, but one suspects that companies prefer a little bullying from time to time to the alternative of detailed command and control regulation and litigation.

This approach is also applicable to areas that are already the subject of command and control regulation. Deregulation is a more risky proposition than trying self-regulation as a first strategy in new areas of concern. The worry is that the very act of

removing certain regulations, and announcing that henceforth
this area will be self-regulated, will be interpreted by industry as
evidence that government is not overly concerned about responsible behavior in the area. The best approach for ensuring that
such an interpretation could not be made would be to have a
reverse sunset clause for all deregulatory initiatives. That is, a set
of regulations would be placed in abeyance for, say, five years. If,
during that five years, industry demonstrated good faith in
making self-regulation work, the regulatory agency would
permanently repeal the regulations; if not, they would automa-
tically come back into force after the trial period.

To work, self-regulation often requires guarantees that
certain interests will be given clout within the system of
informal control. Hence, the privelege of self-regulation might
only be granted on condition that mines have a worker-
management safety committee with powers to stop production
and to demand written responses from the company to safety
problems it raises. Or, it might be granted on condition that
mines with more than fifty employees provide safety officers
with these same powers, plus direct access to the chief executive
officer, rather than access only through the mine manager.
Informal social control is only likely to work in situations where
an organization's guardians of safety have the clout to hold sway
over the worshippers of production.

The above case for self-regulation and deregulation may be
qualified, but it is not a half-hearted one. The empirical
understanding of coal mine disasters we put forth in chapter 2
bespeaks the importance of self-regulation. Disasters were often
found to be the result of a pattern of inattention or sloppiness in
safety matters, diffused accountability within the organization,
poor communication or reporting, and inadequate plans.
Detailed government rule making can never adequately cover the
nuances of managerial inadequacy that lie behind poor planning, poor communication, and weak accountability.

But what can government do to foster the competent
management genuinely committed to safety that the data suggest
is necessary for safer mines? In a word, it can nurture this
competence and commitment by giving managers an opportunity to exhibit it, and rewarding its exhibition with greater
autonomy from government intervention. The social responsi-

blity we expect of mangers is most likely to flourish if it is fertilized by increased trust and autonomy, and if it is pruned when the plant becomes diseased. If governments are tough negotiators in demanding that managers be better trained in safety matters in exchange for this increased autonomy, safer mines at lower regulatory cost will be the result. Government cannot write rules that cover the crucial subtleties of effective safety management, but it can create a regulatory environment in which companies see themselves as having something to gain by good safety management and something to lose by a sloppy approach to safety. That something is provided by regulatory escalation. Now, to the next stage of that escalation.

## ENFORCED SELF-REGULATION

In our discussion of self-regulation, it was argued that companies are in many ways capable of more potent enforcement than government. However, although more capable, they are not always more willing. In areas where the financial costs of compliance are enormous, it is naive to expect the voluntarism of self-regulation to work. But compulsion can be achieved without jettisoning all the advantages of self-regulation for detailed government command and control.

Elsewhere I have argued in greater depth for a middle path called "enforced self-regulation" (Braithwaite 1982b). In outline, the proposal is that each company (or mine) be required to write a set of rules to cover a particular area of concern. The rules would then be ratified by the regulatory agency or sent back for modification. These privately written and publicly ratified rules would then have the full force of law; those who violated them could be punished in the same way as are those who violate laws directly written by government. The main advantage of this aspect of the proposal is that it enables punitive enforcement without having universalistic laws that hamper economic efficiency and create red tape by imposing requirements on all firms that are only needed for some. When rules are particularized for each mine or each company, tougher gas rules can be applied to gassy mines and tougher roof control rules to mines with unstable top. Enforced self-regulation is a way of avoiding the twin problems of rules either too bland and platitudinous to be effecitve (e.g., "accounts must be true and fair") or so stringent and detailed as to cause companies to waste money complying with rules unnecessary in their particular situation.

A further element of the enforced self-regulation model is that a company would be required to internalize most enforcement costs by establishing an internal compliance group to monitor observance of the rules and recommend disciplinary action against violators. Should management fail to rectify violations or to act on recommendations for disciplinary action, the director of compliance would be statutorily required to report this fact to the regulatory agency. This provision would make compliance the line of least resistance for corporate offenders, in most circumstances. That is, the cost of yielding to the compliance director would generally be less than the costs of fighting the investigation, prosecution, and adverse publicity that would almost certainly follow rejection of the compliance group's recommendation.

Under enforced self-regulation, the role of the regulatory agency would be: to determine that company rules satisfied all the guidelines set down by government policy; to ensure that the compliance group was independent within the corporate bureaucracy; to audit the performance of the compliance group; to conduct spot inspections to check that compliance units were detecting violations (and as a total check on companies too small to have their own compliance unit); and to launch prosecutions (particularly against companies that subverted their compliance groups). As we will see below, the beauty of particularistic rules is that they make prosecution easier.

I shall not repeat here the detailed rundown of the strengths and weaknesses of enforced self-regulation provided elsewhere (Braithwaite 1982b). In summary, however, enforced self-regulation's advantages are that rules can be simpler, more readily adjustable to rapidly changing business environments, and more comprehensive in their coverage; companies would be more committed to rules they wrote themselves; companies would be saved the confusion and costs of having to follow two sets of rules in the area of concern (the government's and their own); convictions would be easier to obtain; and compliance would become the path of least corporate resistance. On the debit side there are the regulatory costs of approving the particularistic rules; the risk of greater cooptation of regulators by business under particularism; the fear that companies would so write their rules as to avoid the spirit of the law; the practical problems in guaranteeing independence for the internal compliance group; and concern that Western jurisprudence might not be able to accommodate privately written rules being accorded the status of publicly enforceable laws. In this previous work, I suggested that in considering whether enforced self-regulation is the most appropriate regulatory model, the cogency of each of these strengths and weaknesses in relation to the particular problem must be weighed.

Coal mine safety and health regulation is one of the areas where the strengths of enforced self-regulation are striking. It is hardly surprising, therefore, that in many countries mine safety regulation is one of the few areas where one can see an incipient enforced self-regulation model in practice. The appropriateness of enforced self-regulation to mine safety enforcement arises from the fact that all mines are different geologically, as well as in the way they are laid out. This makes writing rules appropriate for all types of mines—deep and shallow, gassy and nongassy, thick-seamed and narrow-seamed, under water and under land, bituminous coal and anthracite—very difficult.

This reality led to particularistic rule making early in the history of coal mine safety enforcement. Following the recommendation of a Select Committee of the House of Lords, in 1849, that mines in Britain were "too various in their conditions to permit the framing of any safety requirements which could be applied generally," the 1855 Act for the Inspection of Coal Mines imposed only seven general rules and required each colliery to establish its own code of special rules, subject to the approval of a Principal Secretary of State. Because of the weak enforcement of the act, the requirement to write special rules for each mine was widely ignored. However, this was the beginning of a tradition of special rule writing that has been followed to this day in mine safety laws throughout the English-speaking world. The New South Wales Coal Mines Regulation Act of 1982 requires the manager of a mine to write and have approved transport rules, support rules, "schemes for the testing of electrical or mechanical apparatus," and tipping rules for waste. Moreover, the act empowers the minister to require individual mines to write particularistic rules on almost any health and safety issue:

104. (1) Regulations may be made requiring the manager of a mine to make rules or prepare a scheme in relation to any subject-matter which—

> (a) concerns the safety, health, conduct or discipline of persons in mines; and
> (b) is prescribed for the purposes of this section, not being a subject-matter in respect of which rules are required to be made or a scheme prepared under any other provision of this Act.

To leave no ambiguity about how much latitude there is concerning the particularism-universalism of regulations, the act further empowers the writing of regulations that make "the same provision for all cases, a different provision for different cases or classes of cases, or different provisions as respects the same cases or classes of cases for different purposes of this Act" (s.174 [3] [b] [i]). Penalties of imprisonment up to twelve months are provided for violations of these privately written mine-by-mine rules.

The U.S. Mine Safety and Health Act of 1977 provides:

Upon petition by the operator or the representative of miners, the Secretary may modify the application of any mandatory safety standard to a coal or other mine if the Secretary determines that an alternative method of achieving the result of such standard exists which will at all times guarantee no less than the same measure of protection afforded the miners of such mine by such standard, or that the application of such standard to such mine will result in a diminution of safety to the miners in such mine (s. 101 [c]).

Since 1977, more than a thousand petitions for modification have been granted by MSHA. In a few instances, civil fines have been assessed against companies that violated the particularistic standards approved under a petition for modification. However, officials believe that citations for such violations are rare because of the companies' commitment to rules they themselves have sought. The program is not without regulatory cost; each petition consumes roughly three person-days for investigation and approval.

The MSHA regulations also require operators to submit their own plans for ventilation, dust control, and roof support, for the agency's approval. Hearing conservation plans can also be required in certain circumstances. In setting the criteria to be followed in approving roof control plans, the regulations separately define standards for seven different types of roof support techniques. Additionally, operators are free to devise their own unique roof control plans, so long as these would not, in the opinion of the district manager of MSHA, result in less protection of miners. The regulations constitute quite an impressive example of how reasonably clear criteria to limit administrative discretion can be drafted in the face of a variety of technologies, the appropriateness of which depends on the circumstances of a particular mine.

Since December 1979, companies have been criminally convicted in several cases that turned in part on deviations from approved roof control plans. In U.S. v. Vanhoose Coal Co., a mine official was sentenced to sixty days' imprisonment for failing to comply with an approved roof control plan. This offense was responsible for a roof fall in which one Vanhoose miner died and another was injured.

If a roof control plan has a specific requirement that roof bolts be no less than six feet apart, then conviction is much easier than under vague and platitudinous laws of some other countries that roof supports be "adequate" to protect the safety of miners. It is like trying to convict accountants for not following "true and fair" accounting principles, when defense lawyers will always be able to drag up some eminent accounting expert to testify that what was done was "true and fair."

The case for particularistic, privately written and publicly ratified laws in coal mine safety regulation is obvious. All mines are different. Any universalistic standard on the spacing of roof supports is needlessly expensive for mines with excellent top, and unconscionably inadequate for mines with unstable top. The other profound strength of enforced self-regulation is that it can focus enforcement on the overall plan for roof control, dust control, equipment maintenance, and the like. In this book we have seen empirically how disasters arise from inadequate plans. We have also seen how mindless enforcement of specific regulations can undermine the integrity of an overall plan. To have a regulatory strategy which first forces management to sit down and think through a coherent plan, and then enforces compliance with that plan, is to have a strategy that comes to grips with what really causes death in coal mines.

The flexibility of allowing companies to frame their own safety rules encourages companies to come forward with needed new safety technologies; it discourages regulatory agencies from stultifying safety technology at the current state of the art.

There are many other areas where mandated plans could replace specific standards. These include lighting plans, rescue plans, roadway dust control plans, and training plans. Increasingly, governments are requiring that operators provide miners with certain types of safety training. Yet clearly, in different mines, workers confront different types of dangers; training ought to be tailored to the particular configuration of risks confronted in a particular mine.

There is no shortage of critics of the way plan approval works in practice in the United States. Many industry people complained to me that MSHA's district managers often fail to take account of varying conditions in different mines—requiring instead the same pet approaches in all types of mines. Another criticism frequently voiced was the the district manager uses his power to reject plans as a de facto rule-making power to get mine operators to follow his whims, thereby sidestepping the stringent consultative burdens Congress requires before new regulations are written. Operators who are deeply concerned about such abuses can, however, appeal against the district manager's edicts. On the other hand, plans have occasionally been approved over the objections of the union local that they were not tough enough. In fairness to MSHA, after the lodging of such complaints, inspectors have been known to be stationed full-time at a mine to monitor implementation of the new plan; after observing the plan in action, approvals have sometimes been withdrawn and changes suggested along the lines advocated by the union.

In the first few years after the present act was introduced in 1969, there were serious backlogs in approving plans at MSHA district offices. Until officials learned what to look for in approving plans, the process was painfully slow. However, today most plan approvals consume no more than a couple of person-days of the agency's time. With dust control plans, the process has become so routinized that about 90 percent of submissions are simply agency-supplied questionnaires completed by the company. Innovative plans, of course, require a lengthy narrative submission as well, and approval of these may consume up to thirty person-days. Plan approval has certainly not turned out to be a bureaucratic nightmare; company representatives hold informal discussions with government officials to ascertain whether a new approach is likely to be acceptable, before formally submitting it.

The enforced self-regulation model requires more than simply public enforcement of privately written rules. It also requires governmentally mandated internal inspection and enforcement of the internally written rules. Incipient manifestations of this aspect of enforced self-regulation have also long been evident in the coal industry. Under U.S. regulations, specially designated miners conduct preshift examinations of the mine to check for hazards, within three hours of the beginning of each shift and before any miner enters working areas. Preshift examiners are required to record violations of mandatory health and safety standards and, in fact, do so regularly. But in practical terms they are not expected systematically to audit the mine operators' compliance with the law. Rather, the goal is to check quickly every working section for serious hazards. Inspectorial practice is to check the violations recorded in the preshift examination book and to cite them if they still exist, but ignore them if rectified. Other internal compliance functions mandated for coal mine operators are onshift examinations, weekly safety inspections of mine working areas, and regular sampling of respiratory dust concentrations.

These mandated internal compliance responsibilities stop short of requiring an internal compliance group with a compliance director who enjoys substantial powers. There would be merit in bringing these fragmented internal enforcement activities under a coordinated compliance group (while still holding line managers accountable for violations, in accord with practice in the safest companies lionized in chapter 3). It would be ideal if line managers were accountable for violations, and compliance officers were accountable for failures to detect and report serious violations. If the enforced self-regulation model is to work well, there would have to be some exemplary prosecutions of compliance directors who failed to report unrectified violations to the regulatory agency. There have, in fact, been some criminal convictions of individuals in the U.S. for making false or misleading entries in record books concerning preshift examinations.

A statutory obligation to report violations to government is not novel in coal mine safety law. Section 54 (2) of the Queensland Coal Mining Act has this ridiculously broad provision: "As soon as practicable after the occurrence of any breach of this Act that has come to his knowledge, he [the manager] shall report the same in writing to the inspector, warden, mining registrar, or Minister...."

Existing practices in coal mine safety regulation show that enforced self-regulation is legally, administratively, and economically feasible. Its application should be extended to other circumstances in which building a consensus over universalistic standards that can keep up with technological change is difficult or impossible. Moreover, the fragmented existing manifestations of the phenomenon should be coordinated into a coherent enforced self-regulation program under the guidance of a compliance director at each mine who has statutory obligations to report violations to management and then, if management fails to rectify them, to government.

When most coal mines were owned by family companies too small to have both personnel available to write rules and others who could be dedicated full-time to enforcing compliance with the rules, enforced self-regulation was not a practical strategy despite its theoretical advantages. However, in the modern era, as ownership becomes increasingly concentrated in the hands of corporate giants, enforced self-regulation should increasingly become the solution to regulatory overload, a solution that avoids the naiveté of trusting self-regulation in areas where economic interests are profound.

## COMMAND REGULATION WITH DISCRETION TO PUNISH

In some circumstances the voluntarism of self-regulation is naive, and even the degree of voluntarism in enforced self-regulation is unworkable. Government command, by writing specific standards and punishing deviations from them, then becomes the necessary regulatory regime. This section considers the normal discretionary punishment of violations; the next section considers a further escalation of formal regulatory assurance of punitiveness to cope with a situation in which every detected offense must be punished.

One reason for preferring government command over enforced self-regulation is the fear that the regulatory process will be coopted by business. Whether regulatory capture is a greater problem under enforced self-regulation or under government command depends very much on the circumstances.

Universalistic rulemaking, it might be argued, draws out broader resistance to the will of business than could be expected of particularistic rulemaking. Ralph Nader or the Friends of the Earth are more likely to organize against a more lax nationwide effluent standard than they are to oppose an effluent permit for one factory. On the other hand, local citizens who would never be activists at a national level might protest effluent standards which allowed discharges into their neighborhood fishing hole (Braithwaite 1982b, p. 1492).

With coal mine safety, one suspects that in general particularism can harness democratic participation more effectively than universalism. Concerning regulation of mine roof control plans, for example, more interest can be expected from the miners who will be covered by a particular roof plan than from any national activism over coal mine roof safety. And in fact, as we have seen, local union activism does from time to time force MSHA to reverse approvals of roof control plans.

However, there are areas where national debate is clearly more appropriate. In setting maximum allowable limits for dust concentrations in coal mines, local miners will be unable to comprehend the epidemiological studies critical to standard setting and consequently will be in danger of being snowed by the expertise of company scientists. National debate would result in a more even contest between capital and labor. Moreover, in this national debate, other interests besides employer and employee—insurance companies, government health officials, university epidemiologists—could be given a say. Standards, such as airborn dust concentrations, pose dangers of cooptation at a local level too immense to be countenanced; we simply do not want a situation whereby local agreements are being negotiated. The maximum allowable airborn dust level should be national, nonnegotiable, and based on science rather than collective bargaining. Any mine that cannot meet the standard should go out of business.

Beyond the dangers of local cooptation, there is the fact that not all areas of regulation are complex, multifaceted, and subject to constant technological change. For example, it would be a waste of everyone's time to require each mine to write particularistic rules about smoking underground. There can be only one acceptable rule—the prohibition of smoking under all circumstances. So, to the extent that everyone writes the same rule, particularism has been a waste of time; to the extent that everyone writes different rules, some have adopted inadequate safeguards.

Command regulation need not necessarily be more punitive than enforced self-regulation; under the latter model, the state can launch a great number of prosecutions for violations of privately written rules. However, in practice, enforced self-regulation should require less punishment because companies, when confronted with a choice of complying or risking the compliance director's report of an unrectified offense to the government, would normally choose compliance as the path of least resistance.

It would be an overstatement, however, to say that while enforced self-regulation uses punishment only indirectly as a lever to lend clout to internal enforcement, command regulation secures compliance by the direct application of punishment. This is because the fear of punishment under command regulation also gives strength to the forces of internal compliance. To illustrate, let me describe what happened when I accompanied one MSHA official on an inspection. The company sent the safety manager around with us, presumably assuming he would defend the interests of the company against unreasonable regulatory demands. Much of the time he did so. But on two occasions he prefaced a remark to the inspector with, "Don't quote me on this"; he then proceeded to point out an unsafe practice.

The safety manager later explained to me that the inspector "was a big help" to him. When management refused to rectify a problem the safety manager had drawn to their attention, he would point the MSHA inspector in the right direction so that the agency would order the change he wanted. Not only would he get his way, but he would get it in circumstances that enabled him to assert that in future he ought to be listened to more promptly, lest the company land in deeper trouble.

Since that experience, I have spoken to many inspectors in the U.S., Britain, and Australia who reported similar experiences as common occurrences. Even some mine managers who felt they should be using more roof supports, but who had had funding to do this rejected by corporate headquarters, asked the government inspector to demand extra supports. They could then tell the head office that, unless the money was spent, the inspector might stop production. Corporate interests are therefore not monolithic. The fear of punishment under command regulation routinely acts as a force to strengthen the hand of constituencies within the organization that are pushing for safer practices. Punitive command regulation therefore complements self-regulation.

Chapter 4 described how punishment imposed by inspectors can inhibit their capacity to persuade compliance. On the face of it, this makes a good case for the imposition by companies of their own internal discipline as an alternative to public punishment. However, company executives have the same problem as government inspectors when they attempt to punish employees. I was told of foremen who had been sent to Coventry by miners after some among them had been reported to management for safety violations. There is evidence that harmony, morale, and good communications are characteristics of coal mines with low accident rates (Saunders, Patterson, and Peay 1976; DeMichiei et at. 1982). Maximum safety performance is made possible when managers and miners work cooperatively as a closely knit team. Yet, internal discipline can break the bonds of cooperation among people who work together. For this reason, a few coal-mining executives told me that they would prefer that government prosecute individual miners who violate safety laws, rather than having to discipline them themselves. This may be a minority view, but the fact that even some hold it is fascinating.

An important advantage of command regulation and public punishment is that it makes it easier for managers to maintain safety discipline among their work force. Thomas Schelling (1974) has argued that managers of large organizations are rarely in a position simply to issue instructions and expect that they will be carried out. Moreoever, in some cases executives can only secure compliance with their instructions when government backs them. Hence, corporate policies that require the wearing of safety helmets or air-filter masks are notoriously hard to enforce throughout the industry; compliance works best when management can say that the government insists upon it (Schelling 1974, p. 86). In short, an advantage of government command and punitive control is that it gives self-regulators the easy way out of enticing compliance by asserting that prosecution by the nasty government people will be the result unless we pull together and comply.

Beyond that, punishment is necessary under command regulation simply to give credibility to the government's commands. In Queensland, where there has not been a coal mine safety prosecution for a decade, one mine safety officer told an inspection during which the official had indicated to management that a roller was missing from a conveyor belt and must be replaced. On the following inspection, the roller was still missing, so the official wrote in the record book at the mine that the roller must be replaced by the next inspection. When I asked the safety officer what the official would do if the roller was still not replaced at the next inspection, the reply was "I suppose he would take a very dim view." If inspectors are to have credibility, in situations like this the manager should be in genuine fear that he will feel the full force of the law. Managers must not only know that government can impose fines and shut down sections, but actually see them doing so regularly.

Even failed attempts to punish can be far more effective than meekly holding prosecutorial powers in reserve (Waldman

1978). An account in the 1962 Annual Report of the Queensland Chief Inspector of Coal Mines illustrates.

> Proceedings were initiated against the manager of one mine for his failure to conduct searches for smoking material. From the point of view of obtaining a conviction this court action was a failure, but from the overall result and effect upon the industry, I am better pleased and confident it was a resounding success for the action highlighted to all sides of officialdom that it is their responsibility to carry out the requirements of the rules of the Coal Mining Acts so as to protect thoughtless employees from their own individual foolishness. That every branch of management has been informed of its responsibilities in this direction by its own committee of administration is sufficient proof that the action produced the desired result.

In summary, command regulation, whereby government writes detailed rules and punishes those who deviate from them, is appropriate: (a) where the stakes involved are so high that trust cannot be placed in voluntarism and (b) where there is concern that enforced self-regulation would pose a danger of local cooptation; or (c) where the costs of enforced self-regulation would exceed the benefits because simple and invariant hazards are to be regulated.

However, if we are to have command regulation, should it be with or without discretion to punish?

## COMMAND REGULATION WITH NONDISCRETIONARY PUNISHMENT

The problem with discretionary punishment is abuse of the discretion. With business regulation, the most widespread concern is that inspectors who have discretion to issue citations will be coopted by business into a nonpunitive stance. It is immensely difficult for regulatory agencies to write enforcement guidelines that limit the discretion of inspectors (Diver 1980). Bureaucrats in the regulatory agency's head office cannot foresee the complexity of the disparate situations inspectors will confront in the field. Moreover, recognizing (as argued in the preceding chapter) that inspectors must be negotiators building a will to comply, it should also be understood that they cannot afford to be law enforcement automatons who administer consistent justice. To be effective, regulatory agencies must put considerable trust in the wisdom of their inspectors and have very flexible enforcement guidelines.

Perhaps the only way to foster this wisdom is through peer review. Improving the abilities of inspectors is like improving the performance of diplomats. It can't be fully achieved through a set of guidelines. In diplomacy, the only way is to have a more senior person serve on a team with a diplomat in training as that trainee engages in diplomacy, and then talk afterwards about how it might have been done differently. Quantitative output measures—such as the average number of citations, "significant and substantial" citations, and withdrawal orders written per inspection-day—can be used as a guide in selecting those inspectors for peer review who are suspected of not being sufficiently punitive. However, to use the volume of citations as in itself a quantitative measure of performance is to ensure a distortion of regulatory priorities.

> It will induce inspectors to concentrate, in their selection of targets to inspect, evidence to examine, and conditions to report, on readily identifiable violations at the cost of less obvious violations. Since readily observable violations are more amenable to self-policing and very frequently less serious in nature, this tendency will produce suboptimal performance. Basing performance estimates on a more refined classification of violation types can mitigate these distortions. But any workable system of categorizing violations will be sufficiently gross to encompass a wide range of social impacts. Furthermore, any strictly quantitative performance indicator may inspire inspectors to generate citations at a rate exceeding the capacity of the prosecutorial system to handle them, creating pressure on prosecutors to dismiss cases or settle them at low value (Diver 1980, p. 296).

Moreover, inspectors who feel their productivity is evaluated according to the number of citations they write may, by overkill on petty citations, destroy the rapport necessary for persuasion. While quantitative measures should not be used directly to evaluate the performance of inspectors, they should certainly be used to pinpoint inspectors whose performance requires auditing by peer review. The same principle ought to apply to whole regulatory agencies; they should be required to report to the legislature on the number of civil and criminal actions brought each year so that agencies can be selected for subcommittee investigation if there is suspicion of insufficient punitiveness.

It follows from the foregoing, however, that at the end of the day, there can never be satisfactory accountability for the exercise of inspectorial discretion. This is one reason why some are attracted to the MSHA status quo of automatic nondiscretionary citation and sanctioning of mandatory health and safety standards. The question is, however, whether a limitation of discretion is achieved at the expense of a limitation of regulatory effectiveness.

One consequence of nondiscretionary citation at MSHA, as forecast by Diver in the quotation above, has been that penalties have been set very very low to discourage operators from clogging up the prosecutorial system by contesting them. It was seen in chapter 4 that the resulting fines do not deter. The deterrent functions—and more importantly the symbolizing of harm functions—of law enforcement would be better achieved by fewer cases with heavier sanctions.

MSHA inspectors to whom I spoke indicated that they did not, and could not, follow the legislative edict of nondiscretionary citation of every violation they observe. If they did, I was told, they would spend all their time writing citations in the first section of a mine they entered. Just writing a single citation can take twenty minutes—without counting the hours of argument over the citation that might follow, at close-out and health and safety conferences, and during appeals. In practice, then, inspectors try to achieve wider coverage of a mine and issue only informal warnings on minor violations. The fundamental criticism of the across-the-board, nondiscretionary citation is that it can never work in practice. And to the extent that it could approach implementation in practice, it would mean that inspectors would spend most of their time standing still in one section, writing citations.

A counter to the latter criticism is that, if the inspector spends all his time in only one section citing every violation in sight, then other sections, fearing that this might happen to them next time, will ensure that every little problem is put right. One thing seems clear: the optimum strategy must lie some-

where between achieving, on the one hand, minimum coverage
of a mine and maximum intensity of enforcement by writing
every violation seen in a single section, and, on the other hand,
maximum coverage and minimum intensity of enforcement by
walking through the entire mine without writing any citations.
It will be suggested below that this optimum might be
approximated by mandatory citation and punishment of knowing, significant and substantial violations (as defined before
1981) and discretionary citation of other violations.

Mandatory citation across-the-board is also undesirable
because it is sound inspectorial policy to follow a general
guideline of not prosecuting when the offender voluntarily
draws the attention of the inspector to the offense. Inspectors
must encourage openness about compliance problems. This is
suggested as a general guideline rather than an absolute one
because, again, we must trust the wisdom of the inspector to
judge when there are public interests that override the guideline;
we must do this if only to avoid the impression that, no matter
how heinous one's crime, all one need do to avoid punishment
is own up. The point is, nevertheless, that mandatory punishment across-the-board makes impossible even a general policy of
non-prosecution for those who confess.

One of the perverse views I hold about punishing mine
safety offenses is that after a disaster it is enough to name the
individuals responsible in a public report; to prosecute them
subsequently is overkill. There have been a number of cases of
suicide and attempted suicide by persons held responsible for
coal mine disasters and even a case of a man who attempted to
murder his wife and children in a suicide pact. In small mining
communities it is hard for city dwellers to imagine the stigma
that attaches to those perceived as responsible for disasters.
Sometimes even innocent persons suffer from this small-town
stigma. For example, after the Aberfan disaster, in which a waste
tip owned by the National Coal Board washed down a valley
killing 144 people, the tribunal reported that "the tipping gang
and its charge-hand (Mr. Leslie Davies) have all been [wrongly]
bitterly reviled in Aberfan and treated as pariahs." When people
are responsible and are labeled as such, the informal stigma that
arises following mining disasters is quite sufficient to achieve

the deterrence and symbolization of harm that public policy requires.

The real need is to use punishment to deter and stigmatize hazardous conduct that, fortuitously, does not cause a disaster. In such circumstances, the legal finding of fault can be a (rather feeble) substitute for the death and destruction that gives rise to spontaneous stigma. This view is absolutely contrary to the U.S. Justice Department's policy of refusing to proceed criminally for a violation of the Mine Safety and Health Act unless there is a fatality. However, if my view is right, that we can save most lives by using limited prosecutorial resources on violations which do not cause fatalities, that we should forego the opportunity to convict others who are already steeped in blood and stigma, then discretionary rather than nondiscretionary punishment is implied.

A key argument for nondiscretionary punishment, advanced to me by the United Mine Workers of America, is that it takes the pressure off mediocre inspectors who can be snowed by mine managers who know more about mining than they do. One wonders if it is perhaps a good thing that inspectors are swayed by managers when the latter really are more knowledgeable. It is better to give weak-kneed inspectors backbone through audit and peer review of the inspectors who generate low citation rates.

The more important argument is surely, again, that of cooptation. Inspectors are only human; they want to be liked, and they sometimes need to keep open their employment options with mine operators. So, sometimes they will allow themselves to be lulled into nonpunitiveness when they should be punitive. At the same time, an agency commitment to an irreducible minimum of nondiscretionary punishment can enhance the "sense of mission" that motivates inspectors in their unpleasant task (Shover, Clelland, and Lynxwiler 1982, p. 318). These arguments are powerful enough to justify mandatory citation and punishment for knowing offenses that are "significant and substantial." If an offense is knowingly perpetrated when it poses some significant (as opposed to speculative or remote) threat to human life or limb, enforcement should be non-negotiable, nondiscretionary. For such offenses, even the slightest risk of inspectorial cooptation is unconscionable; even a reasoned appeal to negotiate away compliance for other concessions should be rejected. A clear nonnegotiable line should be drawn somewhere, and it is difficult to suggest a better line than the one drawn in defining knowing "significant and substantial" violations, prior to the ambiguous narrowing of that definition in Secretary of Labor v. National Gypsum.

## The Enforcement Pyramid

We have seen that it is advantageous to have the capability of escalating a regulatory strategy from self-regulation, to enforced self-regulation, to command regulation with discretion to punish, to command regulation with nondiscretionary punishment. Equally, once we are plugged into a punitive strategy, it is advantageous to be able to escalate the seriousness of offenses and the severity of sanctions. The primary reason for this is, as we saw in the preceding chapter, that regulating large companies is a negotiating game in which those defending the interests of safety (primarily government inspectors) do not always have more bargaining chips than those defending the interests of production. An enforcement pyramid in which most offenses are at the base, receiving gentle sanctions, and progressively fewer suffer the tougher options, puts offenders in fear of the possibility that they will be among the few who will have the book thrown at them. Equally, it can give them hope that, even though they are guilty of a more serious offense, if they do all the inspector bids by way of reform—and do it quickly—they might receive more lenient treatment. Even when "plea bargaining" of this sort does not really take place, the very fact that officials can and do escalate punishment can generate a suspicion among offenders that it is best to toe the line.

The American Mine Safety and Health Act is the leading example of provision for both a pyramid of seriousness in citations and orders, and, to a lesser extent, a pyramid of severity of associated sanctions. First, for violations of a mandatory health or safety standard, operators (owners) must be held strictly liable for penalties of up to $10,000 for each violation and $1,000 for each day beyond the allowed deadline that the operator does not rectify the violation. If in addition there is evidence that any director, officer, or agent of the operating corporation *knowingly* authorized, ordered, or carried out the violation, such individuals can also suffer civil penalties of up to $10,000 per violation. Further, if such individuals, or the operating corporation itself, *wilfully* violate a mandatory health or safety standard, they may be punished criminally by fines up to $25,000 or imprisonment of a year for a first offense, and $50,000 or five years for subsequent offenses.

While the "knowingly" standard requires proof only on the preponderance of evidence, "wilfullness" must be proved beyond reasonable doubt. "Knowingly" is broadly construed and does not imply bad faith or evil purpose or criminal intent. Moreover, proof of having reason to know meets the standard. A person has reason to know when he has such information as would lead a person exercising reasonable care to acquire knowledge of the fact in question or to infer its existence. A failure to comply is "wilful" if done knowingly or purposely by a person who, having a free will or choice, either intentionally disobeys the standard or recklessly disregards its requirements. It does not require that the defendant intended or foresaw the harm that resulted (Greenspun 1982). The severity of regulatory response can therefore be escalated according to whether the offense is accidental, knowing, or wilful.

The more important pyramid, however, is that of the severity of citations and orders which can be imposed on operators. If an inspector observes a violation of a mandatory health or safety standard, he must cite the violation and set a reasonable time for abatement. Then, if the offender fails to abate the violation within the time allowed, the inspector may escalate enforcement by ordering immediate withdrawal of personnel from areas affected by the violation until such time as it is rectified. When a violation is "of such a nature as could significantly and substantially contribute to the cause and effect of a coal or other mine safety or health hazard," it should be cited as significant and substantial and thereby attract heavier penalties.

The inspector may, in addition, cite the violation as being caused by an "unwarrantable failure" of the operator to comply. The test for an unwarrantable, or unjustified, failure is whether a reasonable person would conclude that the operator intentionally, or knowingly, failed to comply or demonstrated a reckless disregard for the health and safety of miners (*Eastern Associated Coal Corp.*, IBMA 74-18, 3 IBMA 355, 1974). If within ninety days of issuing a significant and substantial citation caused by an unwarrantable failure, another violation caused by an unwarrantable failure is found, the inspector *must* order the withdrawal of personnel from areas affected by the second violation until it is rectified.

In addition to the power to issue a withdrawal order for failure to abate a violation by a deadline, and for a sequence of two different unwarrantable failures, inspectors can order withdrawal when there is an "imminent danger" in a mine, even if not caused by a violation. Withdrawal is, of course, not only a means of protecting miners from danger, but also a far more severe penalty than any fine; enormous loss in production can occur while a section is shut down.

There is a final way that withdrawal can be ordered. The act empowers MSHA to declare a mine as having a pattern of significant and substantial violations. If, within ninety days of an operator's being notified that its mine has been declared to have such a pattern, an inspector discovers a significant and substantial violation, the inspector *must* issue a withdrawal order. If, on the other hand, the mine passes its next inspection without a significant and substantial citation, or has no violations within ninety days, it is then considered to have no pattern.

MSHA has never used its power to declare mines as having a pattern of violations. One reason, MSHA claims, is that it has quite enough powers without it. Undoubtedly, in most cases when a mine is regarded as involved in a pattern of hazardous illegal conduct, a withdrawal could be ordered on the basis of two unwarrantable failures. However, this need not necessarily be possible, and, as the analysis of disaster reports in chapter 2 showed, many tragedies arise from a pattern of minor violations, none of which on its own is outstandingly reprehensible. To quote from the New Sough Wales chief inspector of coal mines, in his report on the Wyee disaster (a report never released because of a government cover-up): "[The roof fall] was caused by a series of small matters, which ignored, will always cause a major accident." This is not only true of coal-mining disasters, but also of product quality and safety-testing disasters in the pharmaceutical industry (Braithwaite 1984), and probably of many other industries as well. The concept of declaring a company as having a pattern of violations is very much in keeping with what actually causes disasters. Hence, it is a pity that the industry for which power to sanction a pattern of conduct has been granted for the first time should be the industry in which such power is least needed because of the existence of other powers that overlap it. The pattern concept owes its birth to a fundamentally correct analysis of the causes of the Scotia disaster, by the U.S. Congress. As Congressman Williams argued in the legislative history of the 1977 Act:

> On our field trips and then at our hearings we looked very closely at the record [of the Scotia mine] and found numerous ventilation citations, we found citations because of inadequate ventilation in that mine, had been issued repeatedly. Inspectors would find inadequate ventilation, would cite the operators. There would be an abatement. A small penalty would follow, it would not be paid for a long time after that, then another inspection would disclose a ventilation violation.... The whole cycle would start again, with no permanent protection being provided to miners....
>
> I feel we have a tool here that will protect miners from repeated violations. We have a method of enforcement that will see a mine closed until the underlying condition is changed, not just a cosmetic abatement for a short period, but a basic reworking of the safety system within the mine (Subcommittee on Labor 1978, p. 1070).

The second reason MSHA advances for its failure to implement the will of the Congress is the difficulty of defining a pattern. The act gives all of the discretion to the agency in drafting criteria for a pattern. Moreover, the legislative history shows that Congress intended MSHA to adopt a flexible position regarding what constituted a pattern:

> It is the Committee's intention to grant the Secretary in Section 105 (d) (4) broad discretion in establishing criteria for determining when a pattern of violations exists.
>
> The Secretary's criteria will necessarily have to be broad enough to encompass the varied mining activities within the Act's coverage.... [W]hile the Committee considers that a pattern is more than an isolated violation, pattern does not

necessarily mean a prescribed number of violations of predetermined standards nor does it presuppose any element of intent or state of mind of the operator. As experience with this provision increases, the Secretary may find it necessary to modify the criteria, and the Committee intends that the Secretary continually evaluate the criteria for this purpose (Subcommittee on Labor 1978, p. 33).

Nevertheless, one would not want to belittle the problems of setting even vague criteria to define a pattern in such a way as to be acceptable to unions and employers. MSHA has experimented with various sets of criteria, none of which was received warmly by any of the interest groups. In addition to the volume of significant and substantial citations, these criteria tended to incorporate qualitative assessments by an inspector of an operator's overall attitude toward compliance as "good," "indifferent," or "poor," and qualitative judgments of the "degree of good faith" an operator evidenced in diligently and rapidly abating violations.

One informant within the agency told me that their experimental criteria were a problem because they would not have resulted in the Scotia mine's being judged to have had a pattern of violations prior to its blowing up. Given that the presumed pattern of violations at Scotia was the raison d'être of the pattern concept, this was a serious concern. The problem surely is that the pattern of violations at Scotia was not a pattern of significant and substantial violations. To be successfully applied, the pattern concept must be taken further than it is in the Mine Safety and Health Act of 1977. A pattern must be definable in terms of minor violations as well as significant and substantial ones. After all, the justification for the pattern concept follows from the empirical understanding that fatalities commonly arise from a chance juxtaposition of many minor acts of negligence rather than from one major one. No better summary of this understanding can be provided than that in the stilted English of the safety policy of IHI, the Japanese shipbuilding giant.

> . . .a large disaster is only occurring by chance out of many small causes which we may overlook, but we will never be able to be free

from disasters if we do not try to eliminate these trifle causes of disaster.

If there were two requirements for a pattern—(1) a minimum number of citations of any sort, plus (2) a consensus among several inspectors that a mine's management had a sloppy attitude toward safety—then it could be used in the spirit Congress intended as a means of putting a stop to the cat-and-mouse of cosmetic abatement; mines could be shut down until the underlying approach to compliance was transformed.

## CONCLUSION

Regulation works best when the regulators can call upon a mix of enforcement strategies and sanctions. It has been argued that there is a place for self-regulation, enforced self-regulation, and command regulation with and without discretion to punish in ensuring mine safety. There should be a preference for trying self-regulation before enforced self-regulation, and enforced self-regulation before command regulation, in deciding upon a strategy for a new area of regulation. This preference arises primarily from the fact that regulatory resources are scarce and that the more informal strategies consume fewer resources. A second crucial justification is the philosophy, developed in chapter 4, of building commitment to the law by first trying strategies that appeal to the better nature of people and, only when they fail, resorting to strategies that assume a lack of good faith.

Ironically, it is conceivable that in a comparatively punitive, formal, and litigious regulatory regime such as exists in the United States, deregulating some areas of enforcement to achieve a mix of strategies could give government more clout. This is because, when government is locked into one predominant regulatory strategy, it can do little to make life more difficult for industry. However, when government can and does shift between self-regulatory and command strategies, depending on how industry behaves, industry has an incentive to show good faith, indeed, to impress government by going the extra mile for safety.

Similarly, shifting from an overall regulatory strategy to the means available to handle a particular violation, if operators and managers understand that there is a pyramid of punitive responses to their wrongdoing, ranging from a friendly warning to criminal conviction and closure of the mine, they will have reasons to work hard at putting things right and to change policies to ensure it won't happen again. However, if the result of detection is the petty fine that follows more than 95 percent of the U.S. citations, nothing an operator can do will change the virtually automatic imposition of the penalty.

Even though the Mine Safety and Health Act provides refined scope for escalating regulatory response, the pyramid of enforcement developed by MSHA is too flat. MSHA needs fewer cases of automatic minor fines at the base of the pyramid and should use the resources thereby freed to get more tough criminal sentences and more withdrawal orders. Then operators really will have an incentive to play the game—there will be both the carrot of avoiding a recorded violation and the progressively bigger sticks of civil fines, closures, and criminal convictions.

Regrettably, however, when the Reagan Administration came to power the MSHA enforcement pyramid was made even flatter than it had been. Significant and substantial violations dropped from over 60 percent of all citations to 22 percent by October 1981, and unwarrantable failures dropped from 2 percent of citations to 0.4 percent by mid-1982. However, by mid-1982 a backlash had developed against the scaling down of enforcement, a backlash fuelled by a run of mine disasters. More inspectors were put on and by the second quarter of 1984 the proportion of citations which were "significant and substantial" climbed back to 42 percent and unwarrantable failures returned to over 2 percent of citations. MSHA's criminal program has not been dismantled as some feared it might, though it remains the situation that criminal prosecutions are the ultimate result of fewer than 0.005 percent of citations. In chapter 6 we will see that regulatory escalation and the variability of response can be further enhanced by creative use of a wider range of sanctions.

# 6. How To Punish

Whether punishment will be swift, effective, and appropriate depends very much on the structure of regulatory decision making. An argument of this chapter will be that power over decisions to prosecute is best decentralized to regulatory agencies rather than centralized under the attorney general, and within agencies it is best centralized in the hands of senior executives of the agency rather than decentralized to individual inspectors or their supervisors.

If regulatory agencies—be they concerned with protecting the environment, preventing antitrust offenses, or promoting occupational safety and health—rely exclusively on the courts for sanctioning, then the frequency of punishment will never be sufficient. They will rush into excessively lenient settlements rather than face the delays, frustration, and costs of litigation. When dealing with an area like coal mine health and safety enforcement, which generates some 140,000 known violations a year in the U.S., to expect the courts to handle all offenses is to ask them to swallow more than they can digest. As Judge Hufstedler has pointed out:

> We have tended to treat every case, whatever its genesis and whatever its dimension, as if it warranted meticulous discovery, several bouts of pleading, a pretrial conference, a 12-man jury, full throttle adversary proceedings, and a few reruns.... We can no longer indulge ourselves in those luxurious assumptions (Goldschmid 1972, p. 447).

Administrative imposition of penalties is essential. Sometimes we underestimate the capacity of nonjudicial, but impartial, tribunals to dispense a justice superior to full-blown court determinations. Administrative justice is often fairer and more efficient than judicial justice in areas having a low incidence of issues of law that require judicial interpretation and a need for specialized knowledge of complex data. When we set up an administrative tribunal to impose civil penalties for regulatory violations, fairness can be increased: (a) by taking pressure off court dockets so that more thorough hearings of the remaining cases are possible; and (b) by replacing a system of infrequent litigation and routine, unofficial sanction without hearing (by blitz inspections, for example) with more frequent official sanction after an administrative hearing. The protections in an administrative hearing might be less than in a court law, but still better   than those under the informal sanctions inevitably resorted to when courts become clogged.

A specialized administrative tribunal established for hearing only mine safety cases can also produce greater consistency of punishment than a district court judge who hears a mine safety case very rarely. The issue here is whether consistency of punishment among mine safety violators is more or less important than consistency between mine safety and other types of offenders. My clear impression is that mine operators are more acutely aware of inconsistencies between how they and other mine operators are punished than of differences between the way they and other kinds of offenders (burglars, polluters) are treated. So, from the point of view of building will to comply by convincing offenders that they are punished justly, the kind of consistency that specialized tribunals can achieve is more important.

## THE STRUCTURE OF ADMINISTRATIVE ADJUDICATION

Diver (1979) has come up with a sensible three-tier proposal for regulatory enforcement. The first type of adjudication he suggests is for small-penalty cases (up to perhaps $200). Where the penalty at stake is so small, neither the defendant nor the agency will find it in their interests to suffer litigation (unless the accused does so for no reason other than to discourage the agency from prosecuting in future). For these low-penalty cases, Diver suggests that the agency define offenses as susceptible to direct observation or measurement, and that a strict liability standard be imposed to eliminate disputes over knowledge or intent. Another requirement for ensuring that the cost of the proceeding is appropriate to the size of the sanction would be that the penalty notice would invite the accused to reply in writing only. Under the second type of adjudication (for sanctions up to $2,000), this protection would be upgraded to an opportunity for an oral hearing. The third type is for cases in which the need for individualized decision making is great and resource constraints less severe. As penalties climb, the ability to pay must compromise an automatic tariff that can reasonably be applied to less serious offenses. Moreover, "attainment of the regulatory objective may require a penalty finely tuned to the precise circumstances of the case" (Diver 1979, p. 1501). Defendants must be given not only a right to a hearing, but also the right to call witnesses and review the written file of the case. A written opinion should accompany the decision, and defendants should have an opportunity to appeal that decision to a higher administrative tribunal and, ultimately, to a court. Then, of course, beyond Diver's models of administrative imposition of penalties, there is the fourth model of criminal prosecution before a judge and jury.

The primary reason for Diver's range of models is to ensure that defendants benefit from procedural safeguards that are in proportion to the gravity of the sanction they risk, instead of oscillating between no procedural safeguards under "kick in the pants," informal justice and a full panoply of safeguards under the criminal model. Perceptive readers will also have seized upon the fact that it is another means of facilitating the pyramid of punishment advocated in chapter 5. A pyramid of punishment is impossible if the costs of litigation are so great as to render enforcement a choice between going all the way or doing nothing. Unfortunately, this is precisely the situation with mine safety law in Commonwealth countries: there are no civil penalties; fines can only be imposed under the full procedural safeguards of a criminal court.

But, assuming the existence of Diver's three-tiered system for the imposition of civil penalties, the question then becomes,

Who should control these various stages of adjudication? Under all of Diver's models, the adjudicator should surely be someone outside the agency. In the U.S., however, most adjudication is done de facto by inspectors, health and safety conferences at district offices, and the assessment office within the agency. An operator can appeal a penalty imposed by the agency to an administrative law judge specializing in mine safety disputes. Undoubtedly, these judges become well informed over time in the technical aspects of coal mining. However, adjudication might be better, swifter, and equally impartial if performed by mining engineers who brought relevant training and expertise to the job as well as a less legalistic, more system-oriented, approach that would focus less on the specific violation than on the implications of the violation for the operation of a total safety plan. Needless to say, a panel of mining engineers scattered across the country to conduct penalty hearings could eliminate a considerable amount of the current dazzling proliferation of layers of adjudication—inspector's decision to cite, mine close-out conference, district or subdistrict health and safety conference, assessment office, administrative law judge, Mine Safety and Health Commission, and normal law courts.

## The Decision to Launch Proceedings

Although agency officials should never have the power to adjudicate, they should always have power over the decision to launch proceedings, either administrative or criminal. The status quo is that few regulatory agencies in common-law countries have authority to prosecute criminally without the participation of an independent prosecutorial agent. This is a fundamental reason why U.S. regulatory agencies such as MSHA ignore the apex of their enforcement pyramid by taking the civil penalty route 99.995 percent of the time.

> Regulatory agencies frequently complain that US Attorneys distort enforcement policy in one (or both) of two ways. First, because they have such enormous workloads, they are far more selective in choosing to bring cases to trial and are more liberal in their criteria for settlement. Second, because of their generalist background and usually parochial political orientation, their

selection criteria differ from those used by the agency. They are likely to be far more responsive to local attitudes about the type of violation involved and the specific defendant. Thus, for example, a US Attorney whose district is more concerned about economic development than environmental quality may prosecute an industrial polluter considerably less aggressively than would the EPA (Diver 1980, p. 287-8).

To put Diver's perception in the terms of the policy analysis of this book, we cannot expect regulatory agencies to make their enforcement pyramid effective if the decision to prosecute at the apex of the pyramid is out of their hands and at the mercy of attorneys who perceive these regulatory offenses not as the apex of an enforcement pyramid, but as the bottom of the prosecutorial barrel. Agency attorneys must be capable of launching their own criminal prosecutions; their enforcement strategy must not be distorted by being sifted through layers of approval in justice departments that have no conception of the enforcement strategy, no comprehension of why it might be more important to prosecute violations that cause no harm than violations that cause disasters, indeed, no understanding of why criminal prosecution of so unglamorous an offense as a coal mine safety violation should be necessary at all. Putting layers of approval between a proposal to prosecute and action is the classic strategy of conservative administrators for cutting down prosecutions. Conversely, if we hold that there are too few criminal prosecutions, then eliminating layers of approval is an important part of removing disincentives to criminal punishment.

If the agency should have the discretion to prosecute, who within the agency should exercise that discretion? Certainly not the inspector. It is important that the inspector be buffered from the resentment that arises when what the operator felt would be a $100 civil penalty is prosecuted criminally. The agency should ensure that its front-line person, the inspector, may say, if he wants, "It wasn't my decision to go criminal," or, "I had to do my job and write up the violation, but I put in a good word for you and pleaded with the powers that be not to prosecute." Sometimes the inspector should take this approach to maintain rapport. On other occasions, such as when the operator has openly flouted the inspector's instructions, he may want the operator to consider him responsible for using the stick. In other words, being an inspector should be like being a backbench politician in a Westminster government: you say you opposed ferociously in the party room the unpopular things the government does, while claiming credit for the good things it does. Inspectors are more effective if they can be diplomatic (nay, hypocritical) in this way. Regulatory agencies should be structured so as to make inspectors diplomats rather than punching bags.

There should be a central office in the agency that takes away from inspectors the discretion to prosecute (as opposed to cite), to decide whether the prosecution will target an individual or a corporation or both, to determine that a mine be declared to have a pattern of violations, to decide under enforced self-regulation to put an internal compliance group through the mincer for not doing its job, and so on. Most importantly, if the agency makes an intial assessment of a civil penalty for an offense, it is imperative that the inspector not be responsible for that assessment.

In fact, there were some impressive features of MSHA's pre-Reagan structure of accountability for assessments. Initial assessments were centralized at MSHA's head office in Washington. But if the operator asked for an oral conference to hear pleas in mitigation against the severity of the penalty, it would be held at one of MSHA's nine district offices. Centralized assessment combined with decentralized mitigation is not only a useful check on bureaucratic error, but also a structure that makes Washington the bad guys while the local MSHA people can win popularity from time to time by mitigating Washington's penalties.

There are other reasons why penalty assessment and decisions to prosecute are better under centralized control. Centralization generates greater consistency of treatment, and it ensures that decisions to prosecute are made by people who know whether the agency has the enforcement resources for consistently prosecuting a particular kind of case. The inspector's closeness to the evidence can impair his capacity to judge how persuasive it will appear to an independent tribunal.

Inspectors do not have the legal resources to settle many
questions of legal interpretation in borderline cases; they would
have to resolve these questions were they writing on-the-spot
assessments. Indeed, under a decentralized assessment system,
one would see a tendency for supervisors to back up the
prosecutorial judgments of their inspectors even when they were
wrong in law or discriminatory in practice. Washington, in
contrast, has no hesitation in ruling that an inspector in the field
has made a mistake. Finally, separating inspection from
prosecution and assessment provides a built-in check on the
quality of inspectorial evidence gathering.

### SEPARATING CRIMINAL INVESTIGATORS FROM INSPECTORS

Another respect in which MSHA still has an instructive
separation of powers within the agency is the appointment of
special investigators for criminal investigations. Inevitably,
gathering evidence for a criminal prosecution is a more abrasive
process, more calculated to rupture rapport, than routine
inspection and citation. Thus, it is a good idea to bring in
special investigators to do the dirty work—extracting confes-
sions, enticing witnesses to turn state's evidence, reading
Miranda warnings. Not only does this protect the regular
inspector from being perceived as the Gestapo, it also ensures
that the demanding requirements of evidence gathering for a
criminal trial are handled by people with police training.

### SUPER-REGULATORY AGENCIES

Centralized control over prosecution, *criminal* investigation, and initial adminstrative assessment of penalties are fine
for an agency as large as MSHA. But every other mine safety
regulatory agency in the world is a tiny fraction of MSHA's size.
In small regulatory agencies, the inspector is, for practical
purposes, part of the head office. There is little prospect of his
nurturing cooperation by asserting that it was the bureaucrats at
headquarters who decided to throw the book at the offender.
Making this possible would be a major advantage of bringing
tiny regulatory agencies under the umbrella of super-regulatory

agencies. In Britain, we saw this with the creation of the Health and Safety Executive to combine in one organization a number of inspectorates with responsibilities ranging from nuclear safety to air pollution to occupational health. Such an agency can have a criminal investigation team of prosecutors and police-trained investigators to select serious offenses reported by the inspectorates for laying criminal charges. The other justification for these super-regulatory agencies is that, with a wide range of powers, they have more clout, more bargaining chips, in the game of negotiation that constitutes most modern regulation (Braithwaite 1984, pp. 376-83).

In short, by combining small regulatory agencies, an unwillingness to breach cooperative relationships through the criminal process can be overcome by having a specialized criminal investigation unit that can afford to be unpopular. While other efficiencies might follow such a combination—for instance, in training inspectors, having a support structure for more decentralized offices, being able to provide central auditing of inspector citation rates, and so on—there are also potential inefficiencies. There is always the fear with a superagency like the British Health and Safety Executive that, for example, the same noise standard that coal mines can easily meet and afford would shut down textile factories. The particularization of standards under enforced self-regulation is one solution to this problem. But beyond that, there is a need for super-regulatory agencies to tolerate the different enforcement strategies appropriate to the different regulatory spheres. As argued in chapter 4, for instance, the infrequent contact between factory inspectors and factory managers means that there is little risk in breaching relationships of trust and cooperation by a consistently punitive approach—a very different situation than exists with coal mines or nuclear regulation. However, just as private corporate conglomerates accept the need for different strategies for different parts of their enterprises, there is no reason why regulatory conglomerates cannot achieve the same diversity.

## PUNISHING THE INDIVIDUAL OR THE CORPORATION

Most of the world's coal mines are owned by either private or state corporations. The question then becomes who should be punished when a mine safety violation occurs, the individual(s) responsible, the corporation, or both? The United States has gone all the way with corporate liability, while in Australia it is individuals and never corporations who are prosecuted for mine safety offenses. It will be argued that the situation in neither country is satisfactory in this sense, and that a judicious mix of both individual and corporate prosecutions is needed.

Under the laws of all coal-mining countries, it is possible to prosecute individual employees for violations of mine safety laws or to prosecute the corporate owner of the mine. Indeed, corporations can normally be punished for offenses perpetrated by individual employees, even when those individuals were acting against corporate policy and company instructions in committing the offenses. However, the extent to which the liability of a corporation approaches strict corporate liability varies from country to country, depending on the defenses companies can raise against being held responsible for acts of their employees. Our purpose here is not to survey these legal differences. Existing differences in the mix of individual and corporate punishment are far less a reflection of legal variation than of prosecutorial preference for individual or corporate targets.

The tradition of individual liability for coal mine safety violations in Australia has a notably unique form. Most prosecutions for mine safety in New South Wales have historically been of individuals, launched not by the state but by the individual's employer. Parnell (1981, p. 35) reports that between 1897 and 1965 there was an average of four prosecutions per year in New South Wales by the government (against owners, agents, managers, or miners), while there was an average of twenty-four prosecutions per year by owners against miners. Up until the last few years, the system worked by government inspectors putting pressure on mine managers to prosecute by threatening the managers and the company with prosecution unless they themselves located and instituted proceedings against a responsible employee. For the state, it was a very cost-effective way of getting prosecutions because, as shown in the preceding chapter, management has intelligence and investigative capabilities for pinpointing responsibility superior to those of government inspectors. Equally, however, it is a system that encourages the absolution of corporate responsibility by sacrificing an individual scapegoat. The practice has faded away in recent years after companies, from fear of the industrial relations consequences of company prosecutions, began to refuse to play ball with the Mines Department.

Mine safety law in many countries is set up to facilitate individual prosecutions by nominating certain individuals as responsible for certain things. Belgian law requires each mine to nominate an "agent responsable" for compliance with mine safety regulations. In Japan, each mine must have a safety supervisor, a technical safety manager, and a safety inspector, all with clearly defined safety responsibilities. The British Mines and Quarries Act of 1954 defines separate duties for owners, managers, undermangers, surveyors, and deputies. The Queensland Coal Mining Act defines the responsibilities of the owners, superintendents, managers, underground foremen, back-shift overmen, deputies, roadsmen, shot-firers, winding engine drivers, boiler attendants, pitheadmen, weighers, pitbottomers, horsekeepers and drivers (!) and "miners and all persons" for underground mines; the act also defines responsibilities in open-cut mines for open-cut examiners, mechanical engineers, groundsmen, open-cut coal mine electricians, and surveyors.

The New South Wales Coal Mines Regulation Act similarly delegates and defines the responsibilities of a number of positions that must be filled at each mine, and the qualifications required of persons who occupy such positions. In addition, responsibilities for certain optional positions are defined (e.g., the mine *must* have a manager and *may* have an undermanager in charge). The act serves to institute a state-imposed management structure. It is no longer the company's prerogative to decide who is responsible for what. Thus, in the 1960s, when many New South Wales mining companies were appointing "overmen" as an extra layer in the management hierarchy between deputies and assistant undermanagers, the Mines Department soon put a stop to the practice, the justification being that these "overmen" had been given production responsibilities by the company but no counterbalancing safety liabilities by the state. The statute not only defines the structure of management but also forbids short-circuiting the structure; the "owner" cannot directly order an employee who is a miner to do anything "otherwise than through, or with the consent of, the manager."

A perennial problem in dealing with corporate crime is that a complex corporate activity gives a picture of confused accountability for any wrongdoing. Everyone can blame someone else. This is more difficult under these coal-mining statutes whereby individual responsibility is more or less clearly defined in the law.

There are other provisions in the New South Wales Coal Mines Regulation Act that can add clarity to accountability. For example, any middle manager or employee who is given an instruction "by or on behalf of the owner" may request confirmation in writing of the instruction from the higher management person who issued it, and such a person must acquiesce to the request. Hence, if a manager is under pressure from corporate headquarters to cut expenditures on safety, he may request confirmation of such a suggestion in writing to make it clear who is responsible for any deterioration in safety at the mine. Indeed, when an instruction is given to a manager or any other employee that the manager believes would impede safety or health, the manager has a *duty* to prevent execution of the instruction until it is confirmed in writing (Section 54 (1)). This amounts to the law's imposing a duty to put the heads of senior executives who compromise safety on the chopping block.

The nominated accountability of such coal mine safety legislation has tremendous advantages. When a manager knows that he has certain statutory obligations, he knows he has a greater risk of conviction for failure to meet those obligations than he would if the state had to prove that he, rather than some other manager, was responsible.

However, when safety management structure dictates the overall management struture, this can breed inefficiency and inflexibility. The law does not change rapidly to accommodate the new management structures demanded by new economic and technological realities. Hence, the huge open-cut coal mines that commenced during the 1970s in central Queensland were saddled with a management structure fundamentally designed for smaller underground mines, which had constituted the entire Queensland coal industry until that time. For example, at the Utah Company's massive central Queensland open-cuts, the manager does planning, marketing, and budgeting and spends only a tiny fraction of his time at the coal face. Utah has therefore delegated a registered manager, a different person from the actual corporate manager, to assume the responsibilities of "the manager" under the act. Utah's position is that only the registered manager spends enough time in the mine workings to fulfill the statutory duties of a manager to monitor safety. In at least one case, a Utah corporate manager does not have a certificate of competency to be a registered manager.

There are two possible interpretations of the Utah strategy of having an actual manager and a registered manager. First, it could be interpreted as a means for the actual manager to protect himself from liability for things that ought to be his responsibility. In an earlier work, I discovered a similar phenomenon in pharmaceutical companies, which had a "vice president responsible for going to jail"; this scapegoat was promoted to a position with formal responsibility for all the matters that could land the chief executive in hot water. Second, the Utah strategy could reflect a genuine need, with operations churning out millions of tons of coal a year, for a person at the helm who is not a technocrat but a strategist with eyes fixed on new opportunities on the company's horizons rather than fixed on the coal face. Both interpretations probably have more than a grain of truth, and each illustrates a weakness of state-mandated management structures: (a) managers who want to find a way around their responsibilities can sometimes do so, and (b) when safety considerations totally dictate the management structure, they can hamstring efficiency by rendering management incapable of keeping in step with changing business environments. State-imposed management structures stultify managerial diversity and inhibit the innovations in management structure that are the source of progress in the efficient running of companies.

One way to solve these problems while preserving the enormously desirable features of nominated accountability is to apply the principles of enforced self-regulation to such accountability for safety in coal mines. That is, the statute would require the company either to adopt a mandatory structure or to submit a management structure to the regulatory agency, setting out in considerable detail who would be responsible for what (Braithwaite and Fisse 1984). The regulatory agency would then either approve the management structure as adequately defining accountability for any violations that might transpire, or send it back for redrafting to achieve greater clarity.

Regardless of the success of such strategies for putting individuals on notice before the event that they will be called to account should certain violations occur, there would still be an important place for corporate responsibility. Granted, individuals are easier to punish effectively than are corporations: they can be imprisoned, corporations cannot; fines can be large enough to hurt them, while fines are rarely sufficient to sting large corporations, which may even pass them on in higher prices; individuals can lose their livelihood through having their certificate (e.g., as a mine manager) revoked, while it is not politically practicable to revoke the mining rights of companies. On the other hand, such a sanction as a community service order can effect far superior restitution, and even deterrence, if imposed on a corporation rather than on an individual (Fisse 1981); although stigma can be counterproductive in labeling individual offenders, it can be an impetus to great reform when directed at corporate offenders (Fisse and Braithwaite 1983); and most importantly of all, although rehabilitation is a dismal doctrine for preventing individual crime, rehabilitation is often easy with corporate offenders (Braithwaite and Geis 1982).

Regardless of how clearly lines of individual accountability are defined, the ineluctable reality of much corporate crime will remain: a number of individuals will each be a part of a whole that no one of them fully understands. Often the system or the plan will be at fault, as seen in the disasters study (chapter 2), and often the best way to get the system changed is to direct pressure on the corporation as a corporation, rather than on any individual within it. This is true whether we are considering U.S. Steel or a state-owned enterprise like the National Coal Board or Charbonages de France. The law should make it clear that, in addition to identifiable individuals having responsibilities, the corporate owner, too, must have responsibilities that transcend those of the transient individuals who pass through positions on the corporation's board and in its executive offices. The law should never encourage the abrogation of corporate responsibility. Corporations, after all, need very little encouragement to salvage the corporate image by passing off their responsibilities onto individual scapegoats for collective wrongs.

Unfortunately, we must show how the same New South Wales Coal Mines Regulation Act that so nicely puts individuals on notice as to their responsibilities also puts corporations on notice that they would be wise to pass off their responsibilities onto individuals.

All mine owners in New South Wales are large companies or state enterprises. The act makes it a defense in any proceedings against the owner for a violation of the act if the owner proves that "He was not in the habit of taking, and did not in respect of the matters in question take, any part in the management of the mine" (Section 164 (1) (a)). It is also a defense if the owner proves that the offense was committed without the owner's "knowledge, consent, or connivance." In other words, the act encourages the corporation to absolve itself of liability by refraining from interfering in the management of the mine. If the company appoints a manager, gives that manager the resources to do the job, and makes no attempt to influence his managerial judgments, then the manager can be hung for any offense, but the corporation is protected. If the corporation is willfully blind to the recklessness of a manager who produces a lot of coal by cutting corners on safety, should the corporation be absolved of liability? Should the law not encourage, nay require, corporations to interfere in the judgments of managers who behave irresponsibly?

The New South Wales law has its origins in both the notion that accountability should be so structured that the mine manager is the captain of the ship, and the notion that professional managers should be protected from the intervention of unprofessional owners. This attitude was undoubtedly sensible in nineteenth-century Britain when the pattern for this legislation was set. At that time, in both Britain and Australia, the problem was perceived as one of the rapacious mine owner who, knowing nothing about mining, would exploit children in their mines, refuse managers the monies to purchase sufficient timbers for roof support, and generally impose a financial squeeze that made safe operation impossible for the manager. The manager, in contrast, was seen as more professional, more attuned to the hardship of the miners he saw every day; and therefore, the thrust of the legislation was to withhold authority from the owner and assign it to the manager.

This conception of the owner-manager relationship is inappropriate to the late twentieth century. We saw in chapter 3 that the mines with the best safety records are owned by large companies that actively monitor the performance of their mines. With good reason, the thrust of business regulation today is increasingly to encourage corporations to be responsible for the misdeeds of their subsidiaries and operating units. We expect reputable companies to audit subunits for compliance with the law, to intervene and make heads roll when a corporate standard of ethical conduct is not met, and to attempt to pull up the safety standards of their weakest subunits to those of their strongest. But a coal-mining company in New South Wales that does all of these things eliminates its defenses against liability for any mine safety violation. The company's way of avoiding liability is to say to the manager: "Here's the tonnage target we expect you to achieve. Achieve it and don't tell us if you have to cut corners to do it." A lesson of the present research is that this kind of corporate philosophy kills miners.

In practical terms, one suspects these issues of corporate exposure to liability produce little concern among coal-mining companies in New South Wales because the government never prosecutes corporations for coal mine safety offenses. The point is, however, that as soon as the government does begin a program of corporate prosecutions, it will find that the existing legislation creates disincentives for corporate social responsibility.

In summary, a strength of the nominated accountability under the New South Wales Coal Mines Regulation Act is that it facilitates individual liabilty, but at the expense of creating a climate averse to the imposition of corporate liability. There is no reason why nominated accountability need necessarily work this way. Just as the act delegates individuals as responsible for the performance of various duties, corporate owners can be required to perform duties ranging from maintenance of a system for monitoring safety statistics, to disciplining mine mangers who fail to run their mines in a safety-conscious fashion, to preparing plans to deal with contingencies such as fires. In the current New South Wales Act, the duties of owners cover seven lines whereas those of employees cover seventeen pages. Moreoever, the owners' defenses of not having knowledge and not having meddled in the management of the mine could easily be eliminated from the act. Thus, there is no reason why a statute that provides for clearly defined individual responsibilities cannot also provide for equally clear corporate duties.

The Australian philosophy of never prosecuting corporations deserves total condemnation. Convicting corporations is the most dramatic way possible of symbolizing the fact that corporations, as corporations, in addition to their individual officers, have responsibilities to be so structured as to minimize safety violations. The symbolic, habit-forming function of law cannot be underestimated: most of us obey the law most of the time because we feel we have a responsibility to obey; a major reason we feel a responsibility to obey is that we see others from time to time being punished and disapproved of for wrongdoing. This is true whether the "us" and the "them" punished are individuals or corporations. Punishing corporations symbolizes the responsibilities of corporations; punishing individuals symbolizes the responsibilities of individuals.

It follows that the American neglect of individual punishment is as bad as the rejection of corporate punishment that we see in Australia, Britain, France, and all the civil law coal-mining countries. We need both. We need a law that reflects the reality that mine safety violations can only be understood as a dialectical interplay between individual and corporate responsibility. There are offenses for which an individual is totally to blame and for which there is no corporate fault. Equally, there are "structural crimes" in which many individuals play small parts, without any single individual's being criminally responsible (*Yale Law Journal* 1979). However, for most mine safety violations, including those involved in most of the disasters described in chapter 2, there is both individual and corporate fault; therefore, both types of sanctioning can and should be brought into play.

## NATURE OF THE PUNISHMENT

Throughout, this book has emphasized the necessity of encouraging informal sanctioning if the limited resources available for coal mine safety are to be maximally effective in saving lives. This informal sanctioning can take many forms: extracting agreement from a company to demote an individual responsible for a violation; issuing a press release indicating the name of the company with the worst accident record or violation rate in the industry over the previous year; warning a deputy that his irresponsibility will entail placing a letter in his file for use against him in any future action to withdrawn his certificate; and so on.

But when formal punishments are to be imposed, what form should they take? At present the only sanctions provided for in mine safety legislation around the world are imprisonment, fines, withdrawal orders, and revocation of licenses for companies or certificates of competence for individuals. The latter two, perhaps because they are such potent sanctions as to seem excessive for most offenses, are very rarely used. Withdrawal orders, which are both potent deterrents and incapacitative measures, are also used relatively infrequently. Fining of both individuals and corporations accounts for the overwhelming majority of formal punishments imposed for coal mine safety offenses.

If fines are sufficiently costly, they may be effective deterrents when directed at individuals, but for corporate targets they have severe limitations. We have seen that in the United States, fines are not heavy enough to deter the bigger coal-mining companies and are generally passed on in higher energy costs to the community as a de facto royalty on the cost of each ton of coal. This, of course, is when the company bothers to pay the fine. We have seen that, often, in the correct belief that the cost for the government of collecting the fine will be greater than its value, American companies simply neglect to pay. In other countries this problem would be easily dealt with by cancelling the company's license to mine coal until they paid up, but in the U.S. only the states have this power, while it is the federal government that does most of the fining. If fines were fewer in number but greater in severity, as argued earlier, they would be an extraordinary cost rather than a royalty, an occasion for making operators think about their conduct rather than a routine event perceived as unavoidable, and the value of the fine would justify the costs of enforcing collection by garnisheeing money from bank accounts, and the like.

Companies are also more likely to take notice of punishments when they are varied in nature rather than routine fines. There is a growing literature on the advantages of having a range of sanctions available to impose on corporate offenders (Clinard and Yeager 1980, pp. 317-22; Orland 1980; Stone 1975; Fisse and Braithwaite 1984; Braithwaite 1984, chap. 9; Criminal Law and Penal Methods Reform Committee 1977; McAdams 1977). Beyond stating that fines have not proven to be an effective sanction and that we need to experiment with different sanctions, it is not within the scope of this book to evaluate systematically the advantages and disadvantages of different penalties. Some are better deterrents than others, some are better at fostering organizational rehabilitation, some have advantages in providing restitution to those who suffer from the offense, while the greatest strength of others is that they preclude the offender's committing further violations. In any case, it is not really a question of one type of sanction's being better than another. To suggest that one kind of corporate sanction is superior to others is like suggesting that fighters are better than bombers for an effective defense force. In the battle against corporate crime we need an arsenal that enables us to take a different kind of arrow from our quiver for each corporate target.

For the purposes of this book, therefore, we need to do no more than list the types of corporate sanctions that have been suggested and provide further references containing assessments of the advantages and disadvantages of each.

*Cash Fines.* The company is ordered to pay a dollar amount (Nagel 1979).

*Equity Fines.* The company is ordered to issue new shares to a victim compensation fund. After a 5 percent equity fine, the victim compensation fund will own 5 percent of the shares in the company, and the remaining shareholders will have the value of their holding diluted by 5 percent (Coffee 1981; Fisse and Braithwaite 1984).

*Publicity Orders.* The court orders placement of advertisements or other publicity in mass media outlets notifying or warning the public (or certain publics) of a company's offense (Fisse and Braithwaite 1983).

*Internal Discipline Orders.* The company is ordered to investigate an offense committed on its behalf, to discipline culpable employees, and to report to the court on what it has found and done (Criminal Law and Penal Methods Reform Committee 1977, p. 361).

*Preventive Orders.* The company is ordered to change certain policies or standard operating procedures, expand certain internal compliance activities or budgets, appoint persons to certain new positions with specified authority to prevent future offending (Solomon and Nowak, 1980; Criminal Law and Penal Methods Reform Committee 1977, p. 359; Stone 1975, pp. 186-98).

*Corporate Probation.* A relevant expert is appointed under a corporate probation order to supervise implementation of internal reforms similar to those under preventive orders (Yoder 1978, p. 53; Yale Law Journal 1979).

*Community Service Orders.* The company is ordered to perform as an organization some work of community service relevant to its expertise (e.g., a coal miner testing a new approach to revegetating reclaimed strip-mining land) (Fisse 1981).

A more detailed case for the availability of all these sanctions has been made elsewhere (Fisse and Braithwaite 1984). Suffice it here to make the point that, even as there are many ways we can be punished as individuals, there are various means of punishing a corporation. In light of the following conclusions and those of chapter 5 about the circumstances in which corporations ought to be punished, it would be a pity if readers assumed punishment to mean only fines.

CONCLUSIONS

1. The imposition of both civil penalties for violations of mine safety and health laws under a system of administrative adjudication and criminal penalties in courts of law must be possible.

2. Administrative adjudication should be tiered so that procedural safeguards can vary in proportion to the gravity of the sanction that can be imposed.

3. Regulatory agencies, as opposed to justice departments, should have power over decisions to launch all types of proceedings.

4. Inspectors should only write citations; decisions on whether and how to proceed further should be centralized in the agency.

5. Responsibility for criminal investigation should not rest with ordinary inspectors, but with specialized criminal investigators who come in after the inspector has written his ordinary citation.

6. In jurisdictions with small mine safety enforcement agencies, there is virtue in becoming part of a super-regulatory agency and relying on a central office within this agency for specialized support, such as criminal investigation and prosecution.

7. Both the American preoccupation with corporate liability for mine safety and health offenses and the Australian infatuation with individual liability are suboptimal. The need is for a more mixed strategy of corporate and individual liability.

8. Mine safety and health law should require that certain individuals be delegated as accountable for performing various safety responsibilities in order to minimize diffused accountability throughout the management structure. Certain collectivities should similarly be delegated as accountable for specified safety duties.

9. A disparate arsenal of individual and corporate sanctions is needed to accommodate the variable situations in which punishment is required, to make punishment an event that attracts interest rather than a routine royalty on production, and to help refine the pyramid of enforcement.

# 7. Regulation, Productivity, and Saving Lives

The fundamental dilemma in regulating industrial safety is often viewed as a tradeoff between lives and productivity. Regulations to protect miners often do slow production. It is a misunderstanding, however, to treat the situation as a simple tradeoff. This chapter will argue that fundamentally, though far from invariably, improved productivity is in the interests of health and safety.

But this leads us to the real dilemma. By definition, the higher the productivity, the smaller the number of workers in the mines and exposed to risk for each ton of coal extracted. Safety regulations that reduce productivity are perverse in their impacts: on the one hand, they improve protection for those in the pit; on the other hand, they increase the number of people who must be exposed to the residual risks of the mine. In other words, we must be concerned lest our conclusions about how to mix punishment and persuasion so as to maximize observance of regulations could actually increase the death toll in coal mines when the regulations inhibit productivity.

## DOES REGULATION REDUCE PRODUCTIVITY?

Executives of some companies that owned many mines told me that their internal analyses had shown that their most productive mines were those with the fewest accidents and the fewest citations for serious safety violations. Recent studies by the National Academy of Sciences (1982, p. 98), the General Accounting Office (1981, p. 44) and DeMichiei et al. (1982, p. 27) have confirmed this on the basis of publicly available U.S.

data. Hopkins (1984) also found, in New South Wales, Australia, that the most productive mines had the lowest accident rates. The most commonly accepted interpretation of these surprising findings is that safety and productivity share the same sire—competent management. As the National Academy of Sciences concluded: "A management that can plan well to increase production can also plan well to improve safety" (1982, p. 15). And, of course, competent managers have an interest in doing both because accidents cost dearly in interrupted production, damage to working areas, and workers' compensation payments or increased insurance premiums. On the other side of the Atlantic, the thesis of fundamental compatibility between safety and production has been equally strong. Collinson, the National Coal Board's chief safety engineer, has argued that "...effective management for safety actually increases productivity and therefore...managing safety and coal productivity are entirely compatible and jointly productive of the required end-result—safer and higher productivity" (1976, p. 3.3).

During my company interviews I was given many examples of how good management had simultaneously reduced accidents and improved productivity. Job safety analyses often reduce accident risks by redesigning tasks to make them simpler—perhaps requiring less clumsy movements or carrying heavy materials about for shorter distances. Clearly, a job that is rendered safer by making it simpler can also be performed more rapidly and efficiently because of that same simplicity. De-Michiei et al. (1982, p. 15) found that, at mines with low accident rates, materials handling was an important factor in mine design and equipment requisition. To take another example, a well-run training program will give miners a clearer understanding of what can go wrong; this understanding can be used to cope with production problems as well as safety breakdowns. A poorly trained miner is both less productive and in greater danger.

Many safety problems result from failure to maintain equipment in permissible condition—frayed electrical cables that form potential ignition sources, machines impregnated with coal dust creating a fire risk, methane detectors and water sprays that become clogged, and so on. But many production stoppages arise from breakdowns in the same or similar equipment. Hence, an effective maintenance program can simultaneously improve safety and productivity. Lilly (1979) found that mines with both scheduled maintenance shifts and that practiced preventive maintenance had higher productivity. Indeed, more highly productive mines were often found to spend as much as five times more time on preventive maintenance than did mines with low productivity.

Methane drainage, whereby small pipes are drilled into a seam to drain methane prior to mining, can reduce the risk of explosion and produce receipts for sale of the gas that generally offset or exceed the cost of extracting it (Breslin and Anderson 1976, p. 113). One could go on and on with examples of how capable managers are forever exploiting programs that are at the same time good for safety and good economics.

Even though the number of areas in which safety and productivity can be simultaneously enhanced by competent management is so great as to cause a negative correlation between mine accident rates and productivity, there remain many other areas in which a choice must be made between profits and people. Consider the requirement imposed by all nations that mines must have two exits so that, if one is blocked, trapped miners can escape through the other. Digging these extra shafts has cost mine owners enormous amounts of money, but it has saved thousands of lives.

The cause célébre of those who wish to highlight how regulation can reduce productivity is the U.S. Federal Mine Safety and Health Act of 1969. We have already seen how, making comparisons at one point in time, safer mines are more productive mines. In general, we get the same picture from time series data: in both Britain and the United States, historical periods when fatality rates dropped rapidly were periods when productivity grew rapidly (Collinson 1976, p. 3.7; Zabetakis 1981; National Academy of Sciences 1982, p. 39). Rather than reflecting any fundamental compatibility between safety and productivity, some would argue that this reflects a relationship whereby, in periods of increasing productivity, people become wealthier, and as they become wealthier, they demand more safety (Viscusi 1983).

A major exception to this trend was the 1970-77 period in the United States—the years following the 1969 act. After its enactment, fatality rates fell dramatically, but productivity also dropped substantially. Prior to the act, productivity had been steadily increasing. The rot stopped in 1978, when productivity again began to rise; indeed, since then it has risen at a faster rate than it had during the 1950s and 60s.

The most significant change in mining practices arising from the 1969 act concerned prohibitions against miners working under unsupported roof. Before the act, it was common for continuous miners to advance 30 meters before being withdrawn from the face to enable the roof to be bolted. By forbidding miners to work under unsupported roof, the act reduced the distance of advance to the length of the machine, about 6 meters. Much time is "wasted" through more frequent shifting out of continuous miners and moving in of roof bolters. The Consolidation Coal Company undertook a simulated time study of the production time it lost as a result of this and other new requirements of the 1969 act. The conclusion was that 38 percent more coal would have been produced in 1977 under prelaw conditions than the company actually produced (Consolidation Coal Co. 1980, p. 62). However, as the General Accounting Office pointed out: "The coal industry has attributed a greater percent of the decline [in productivity] to the act than other research indicates" (1981, p. 36). Gordon et al. (1979, p. 4), for example, found that the most significant aspects of the act caused an average 18 percent decline in productivity (see also the review by Oak Ridge Associated Universities [1979]).

The General Accounting Office (1981, p. ii) concluded that most of the decline in U.S. coal productivity between 1970 and 1978 could not be explained by the 1969 act. Certainly, the act was a "major cause" of declining productivity between 1970 and 1973, but subsequently faded away until, ultimately, it "was no longer a factor in continuing productivity decline" (General Accounting Office 1981, p. ii). Other factors found to be important in causing the decline in productivity during this period were a reduced quality of labor-management relations and 1974 union contract requirements mandating extra helpers on mine-face equipment. In any case it would have been unrealistic to expect a continuation of the quadrupling of productivity that occured between World War II and 1969, since this had been achieved by a process of mechanization, which had been all but completed by 1970, and by the closure of thousands of less efficient mines, which could not compete during the declining demand for coal of the 1950s and early 60s—a factor that disappeared with the rising demand for coal by the 70s.

Even though the adverse impact on productivity of the 1969 act in the U.S. was not as severe as industry claims, there can be no doubt that some significant adverse effect occurred. But even this amount must be put in perspective. It may be that new technology will ultimately eliminate many of these productivity deficits. Robots, or remote-control panels that place operators back from the coal face, will ultimately solve the problem of miners working under unsupported roof, and perhaps within this decade we will see widespread use of fully integrated miner-bolter systems, which will eliminate the production bottleneck of stopping miners while roof supports are put in place (General Accounting Office 1981, p. 88).

After roof control, the second major concern of the industry with respect to the 1969 act was tougher ventilation provisions that require brattice to be kept within 10 feet of the working face. Brattice is flame-resistant curtain hung from the roof to the mine floor to channel air, dust, and methane away from the workface. One large Illinois mine complained that it was obliged to employ four additional brattice workers to comply with the 1969 act (General Accounting Office 1981, p. 47). The U.S. Bureau of Mines is currently funding promising projects to make remote movement of brattice feasible. In other words, many of the productivity setbacks arising from regulation are short-lived—lagging technology soon catches up to the new environmental contingencies. Indeed, regulation can be the mother of inventions that not only negate the productivy loss but positively take productivity to new heights (Ruttenberg 1981). Hence, in any analysis of a supposed tradeoff between productivity and people, it would be a mistake to assume that the productivity impacts of today will be the same as those of tomorrow.

In summary, there are so many ways that safe practices are more productive than unsafe ones that, overall, there is a positive correlation between safety and productivity in the coal industry. Nevertheless, many significant areas remain in which productivity must be sacrificed if safety is to be improved. It is in these areas that controversial safety regulations are to be found. Is the improvement in safety worth the loss of productivity? This is the question we now address. Finally, it must be remembered that the inspections required to enforce all regulations, even those which do not hinder production, themselves cause nontrivial disruption of production (Oak Ridge Associated Universities 1979, p. 243; Hill et al. 1981, pp. 11-12; Boden, Zimmerman, and Spiegelman 1981). Thus, we might also ask, Is it worthwhile to disrupt production to check compliance with regulations that only minimally improve safety in industry?

## LOST PRODUCTION AND LOST LIVES

When legislators pass health and safety laws, they sometimes instruct the regulatory agency to write and enforce regulations to protect human life to the maximum, without taking account of the economic consequences. However, in Britain and Australia, mine safety legislation has always implied that regulation ought to weigh economic realities by requiring compliance with regulations only "as far as is reasonably practicable." Since the late 1970s, in the United States, regulatory agencies have had to prepare "inflation impact statements" for all new regulations. Such policies are little more than a political sleight of hand because they give no real guidance as to the circumstances in which economics ought to outweigh safety. An argument will be presented here for one very explicit kind of guidance that legislatures ought to give on how to trade off the safety and productivity impacts of regulations. That guidance is:

*Any rule or enforcement of a rule should be stopped if the number of lives and injuries it saves is less than the increase in death and injury caused by reduced productivity arising from the regulation.*

This means that, when evaluating the efficacy of a regulation, we ought to ask how much loss of productivity (tons per worker-hour) will result from it. Then we must ask how many extra people must be put down the pits to extract the same amount of coal, and how many of those extra people will be killed and injured during their time in the mine. If that number is greater than the number saved from death and injury by the extra protections afforded by the regulation, then the regulation is indefensible. The proposal, therefore, does not require that dollars be traded against lives; it simply requires that dollars at least be saved in circumstances where the suffering from more widespread exposure to risk is greater than that prevented by the regulation itself. When we realize the fundamental compatibility of safety and productivity in coal mining, we can render morally unproblematic a large proportion of the regulator's decisions in balancing lives and dollars. They remain, however, empirically highly problematic, since productivity impacts of regulations are difficult to ascertain, and because, as pointed out earlier, there may be a technological lag which will ultimately see the elimination of the productivity setbacks from a new regulation.

One may ask whether a decline in productivity might simply mean that less coal would be mined rather than the number of miners increased to make up for lost production. The answer is fairly clearly no. When demand for coal goes up, more coal is mined. When demand drops, less is mined. If more miners are required to produce the coal for which there is a demand at any point in time, then more miners will be employed, new mines and new faces will be opened up. With the declining demand for coal between 1950 and 1969, in large part as a result of railways abandoning coal as a fuel, some 3,000 underground coal mines were shut down in the U.S., and the number of miners dropped from 373,000 to 133,000. By 1980, under the influence of a growing demand for coal by electric utilities and other sources, the number of U.S. coal miners had grown to 245,000. Extra employees were needed in the 1970s because demand was growing and productivity declining. Even for the world's largest exporter of coal, the United States, around 90 percent of the demand for coal is domestic (Zimmerman 1981, p. 7); thus it would be difficult even to sustain the argument that lower U.S. productivity would reduce the amount of America coal mined at the expense of increased production in other countries, thereby shifting the deaths and injuries resulting from reduced productivity offshore. I could be wrong in arguing that reduced productivity means more miners. In some circumstances, lower productivity could sufficiently increase the price of coal that users would switch to other fuels. If consumers do switch, the only realistic candidates to replace coal in the medium term are nuclear power and oil (Zimmerman 1981), which have their own enormous safety problems (e.g., Carson 1981). The weight of opinion from policy analysts favors increasing the use of coal and decreasing dependence on oil and nuclear power.

If reduced productivity means employing more miners to extract the same amount of coal, is this really a bad thing? Surely more employment is desirable. I will spare readers the basic economics lesson on why it does not make long-run economic sense to create new jobs by reducing the productivity of existing jobs. Ultimately, a society cannot afford to pay people who, without benefit to the economy, are in effect employed to dig a hole and then fill it in again. The point here is that it is more than just bad economics to pay someone to dig a hole and then fill it in, if there is a significant risk that the hole will collapse and kill him.

This reasoning leads to the view that mechanization has had, on balance, an overwhelmingly positive effect on coal mining. On the one hand, it has increased the national wealth of coal-mining countries; on the other, it has eliminated the horrific disasters in which hundreds perished—in a mechanized industry there are hardly any mines with hundreds of miners underground at one time. Maybe the reduced slaughter has been purchased at the sorry price of lower employment in the industry, but this unemployment has assuredly been more than offset by the employment growth generated after the wealth from mechanized mines was invested to create jobs elsewhere in the economy.

It follows from the analysis here that I do not totally dismiss as propaganda American coal owners' claims that the 1969 Mine Safety and Health Act in some ways may have made coal mining less rather than more safe. In the battle over statistics, the Mine Safety and Health Administration points to the dramatic drop in fatality rates per 200,000 hours of exposure that occurred in the years after the act was introduced. For their part, the industry points out that the decline in the rate of fatalities per million tons of coal mined since the act has been more modest (Consolidation Coal Co. 1980, p. 28). Henderson (1976) has shown that (at least for the first three years after the act's introduction), while the act could be shown to have reduced fatalities per million worker-hours, fatalities per million tons did not significantly decrease, because of declining production during those years. Looking back now, we can see that fatalities continued to fall substantially after 1973, and that after 1977 productivity bounced back—so, at the end of the day, the 1969 act was a success on any yardstick.

Essentially, the industry is right that the number of people who have to be killed and maimed to produce a million tons of coal is the more important statistic for public policy. It is all very well to show that, if you go down a mine today, you are less likely to be killed than you were in the past, but if more people must go into the mine for each ton of coal produced, we must add the risks they face.

The aspect of the industry's case against the 1969 act that concerns us here is quite simple: if productivity had remained at the 1969 (pre-act) level, the same amount of coal could have been mined between 1970 and 1979 with the loss of 365 fewer lives (Consolidation Coal Co. 1980, p. 87). This projection is based on the (false) assumption that the reduced number of miners who would have been required, had pre-law productivity been maintained, would enjoy the same (lower) fatality rates per 200,000 hours as post-act miners. On the other hand, it might be argued that the assumption is conservative because the new miners who joined the industry as a result of the drop in productivity were predominantly young and inexperienced. The National Academy of Sciences (1982, pp. 100-103) found that miners under 24 years of age have three times the injury rates of those over 45, though age differences in fatality rates were minimal. The estimate of 365 lives lost also rests on an assumption that evidence reviewed earlier has shown to be false—that the post-act decline in productivity was totally or even primarily explicable by the act. Nevertheless, though the 365 figure wildly exaggerates the effect of lost productivity as a result of the act on loss of life, their basic point is undeniable that the drop in productivity caused by the act resulted in *some* loss of life.

There are other reasons why the 365 figure is nonsense. One way by which the 1969 act reduced productivity was to increase the burden of paperwork on mine opertors. About 5 points of the 45 percent drop in production per worker-hour between 1969 and 1977 were produced by the appointment of additional office workers (General Accounting Office 1981, p. 38). These extra office workers are clearly not subjected to risks of fatality equal to those below ground.

The act was also said to have disrupted productivity as a result of a steep increase in the number of withdrawal orders issued after its introduction. Inspectors are empowered to issue orders that miners be withdrawn from a section of a mine whenever dangers in that section are excessive. In the first year under the new act, withdrawal orders increased from 1,493, in 1970, to 4,390, in 1971. Because labor is rendered idle, withdrawal orders reduce production per worker-hour. However, they certainly do not have the effect of increasing accidents by leaving more miners exposed to risk—in fact, quite the contrary.

Clearly, we must look at what lies behind a drop in productivity caused by regulation before we can decide whether the regulation produces a net benefit or decrement to safety. If a regulation only decreases productivity by increasing the number of clerical workers on the surface, then we can forget about increased exposure; the new regulation can be evaluated simply on the basis of whether safety is improved for those already underground. If, at the other extreme, a new regulation requires that an extra person be employed on each roof-bolting machine at the face, then we are putting new people into areas of maximum risk. We must balance the deaths and injuries the new roof bolters will suffer against the greater safety all miners who walk under the roof might enjoy. Between these extremes, if we employ more miners on rock-dusting tunnels away from the working face to stop the spread of any possible explosion, we are deploying extra employees at areas of low to intermediate levels of danger. Their exposure to this intermediate danger must be weighed against the reduced risks to all in the mine of dying in an explosion.

Conclusion

In short, new regulations or inspection programs must be looked at one by one. We must ask, first, how many accidents would be avoided for those already in mines as a result of the regulation or program. Then, we must ask how many accidents are likely to befall additional people who are sent into the pits as a result of the regulation or program. If the latter exceeds the former, the regulation or program should be abandoned. The effect of such a decision process would be to decide *on the basis of safety* to eliminate a good number of the regulations that hinder productivity. A major concession would be made to the position that economics ought to be considered in regulation. This would be done without endangering lives—indeed, by saving them. It would be a principled injection of economics into safety regulation, which would define a clear rationale for taking productivity impacts into account—one that involves difficult empirical questions about the safety and productivity impacts of regulation, but at least one that is clearly defined. Such a concession would be a major advance in clarity of public purpose over the meaningless injunction that regulatory agencies enforce regulations "as far as is reasonably practicable."

This proposal for considering the economics of regulation before charging ahead with the mix of punishment and persuasion recommended in chapters 5 and 6 does not go as far as mine owners would like. It provides a principled rationale for considering the productivity impacts of regulation, but not the impacts on costs other than labor. For example, a regulation requiring that new roof bolts be made of tougher, more expensive steel could never be rejected under the rationale presented here, because it would increase costs without reducing productivity. This proposal therefore leaves us without a solution to a great many of the morally perplexing tradeoffs between dollars and safety.

Notwithstanding our failure to resolve the remaining dilemmas of choosing between miners and money, it has at least been shown that regulatory strategy cannot achieve its mission of maximizing safety without considering its productivity impacts.

# 8. Conclusion

A book that seeks to analyze policy in a way relevant to all coal-mining countries cannot draw highly specific conclusions. Readers must fashion the raw material presented herein according to its relevance to the institutional realities of their own societies. Moreover, as we have seen, this must be done with an eye to cultural differences, placing heavier reliance on punishment in some societies (e.g., the United States) than in others (e.g., Japan).

Yet, there is a common basis for some of the broader conclusions drawn in the preceding chapters. In all countries, death in mines could be dramatically reduced if there were greater compliance with their laws. The world over, fatal accidents are attributable to the causes identified in part 1: to defects in planning, in communication, in definition of responsibilities, and in training/supervision; to patterns of inattention concerning hazards; and to the inadequate authority of safety specialists. So, in all countries, there is a case for directing law enforcement at plans as well as at specific rules (poor planning); for imposing duties to know, to report, and to demand that instructions that compromise safety be put in writing (poor communication); for requiring that companies publicly nominate persons responsible for specified safety duties, and provide adequate training and supervision; for declaring mines to have a pattern of violations and thereby exposing them to more potent enforcement; and for insisting that appropriate authority to stop production and order reform be given to safety specialists without compromising the principle that it is line personnel who must be held accountable for offenses.

In no society can wholesale reliance on either punishment or persuasion be sound policy. Each society must find the point of optimum mix that maximizes the synergy between punishment and persuasion. This is the point at which any *more* punishment would so sap the will to comply, so sour relationships and sabotage the inspectorate's capacity to persuade, that behavior would become less rather than more safe. It is the point at which any *less* punishment would so undermine the deterrent, symbolizing of harm, and the rehabilitative and incapacitative functions of punishment (and so weaken the belief of offenders that persuasion is best heeded because punishment is the alternative) that compliance would also worsen.

Compliance is most likely when regulatory response can be escalated, when governments realize that the only way to make best use of scarce enforcement resources is to negotiate a level of regulatory intrusiveness for each hazard that is proportionate to the degree of good faith industry has shown in fostering compliance regarding that hazard. Chapter 5 argued that the most fundamental hierarchy of regulatory response should be from self-regulation, to enforced self-regulation, to command regulation with discretion to punish, to command regulation with nondiscretionary punishment.

Furthermore, there is a need for a pyramid of severity of regulatory orders (e.g., the U.S. pyramid of violations, significant and substantial violations, unwarrantable failures, patterns, withdrawal orders, criminal violations) and an overlapping pyramid of sanctions (e.g., formal warning, civil monetary penalties of different levels, and criminal sanctions of different levels and types up to mine closure). The whole range of punitive response must be used regularly against both individual and corporate offenders. Because the costs of regulatory response increase as we ascend these hierarchies, most regulatory activity must be at the base of the pyramid. Nevertheless, it has been argued that no society uses the weapons at the apex of the pyramid, particularly criminal punishment, sufficiently. All existing mine safety enforcement pyramids are suboptimally flat. In the United States it might be desirable to reduce the number of cases at the lowest level of regulatory response by taking no regulatory action at all and relying on persuasion in these less serious situations; the resources thereby freed might then be redeployed at the apex of the pyramid to ensure that real deterrence and symbolizing of harm are achieved by some severe criminal sentences.

The latter strategy implies that inspectors must be trusted to have wisdom, diplomacy, and incorruptibility in exercising the choice between punishment and persuasion. The only means of ensuring that these virtues are attained by inspectors is auditing and peer review, plus discriminating recruitment policies that will attract highly qualified people.

Finally, it has been argued that a regulatory strategy, whatever its mix of punishment and persuasion, cannot be maximally effective in saving lives if it does not consider its impact on productivity. This is because, when enforcement results in requiring more miners per million tons of coal, more people will be exposed to greater or lesser risks, depending on where the extra personnel are required in the mine. Hence, I submit that law enforcement should cease where it saves a number of lives and injuries that is less than the increase in death and injury caused by law-enforcement-induced productivity loss.

Because, as we have shown, there are many circumstances in which law enforcement can do more harm than good, it follows that imposing philosophies of consistency and equity in punishment, of punishing criminally all those who can be proved to deserve it, will result in greater death and injury in mines than is necesary. Equally, it has been shown that criminal punishment is a powerful and universally underutilized tool for making mines safer.

# Appendix
# Citations to Mine Disaster Reports

UNITED STATES

Bureau of Mines, Final Report of Major Mine Fire Disaster, No. 22 Mine, Island Creek Coal Company, Pine Creek, W. V., 8 March 1960.
———, Final Report of Major Mine-Explosion Disaster, Viking Mine, Viking Coal Corporation, Terre Haute, Ind., 2 March 1961.
———, Final Report of Major Mine-Explosion Disaster, Mine No. 2, Blue Blaze Coal Company, Herrin, Williamstown County, Ill., 10 January 1962.
———, Final Report of Major Mine-Explosion Disaster, Robena No. 3 Mine, United States Steel Corporation Coal Division, Carmichaels, 6 December 1962.
———, Final Report of Major Mine-Explosion Disaster, Compass No. 2 Mine, Clinchfield Coal Company, Dola, W.V., 25 April 1963.
———, Final Report of Major Mine-Explosion Disaster, No. 2 Mine, Carbon Fuel Company, Helper, Utah, 16 December 1963.
———, Final Report of Major Mine-Explosion Disaster, No. 2a Mine, C. L. Kline Coal Company, (near) Robbins, Tenn., 24 May 1965.
———, Final Report of Major Fire and Explosion Disaster, Mars No. 2 Mine, Clinchfield Coal Company, Wilsonburg, W.V., 16 October 1965.
———, Final Report of Major Mine-Explosion Disaster, Dutch Creek Mine, Mid-Continent Coal and Coke Company, Redstone, Colo., 28 December 1965.
———, Final Report of Major Mine Suffocation Disaster, Dora No. 2 Mine, Doverspike Brothers Inc., Dora, Pa., 1 June 1966.
———, Final Report of Major Mine-Explosion Disaster, Siltix Mine, The New River Company, Mount Hope, W.V., 23 July 1966.
———, Final Report of Major Mine-Explosion Disaster, River Queen Underground Mine No. 1, Peabody Coal Company, Greenville, Ky., 7 August 1968.

———, Official Report of Major Mine Explosion Disaster, Nos. 15 and 16 Mines, Finley Coal Company, Hyden, Ky., 30 December 1970.

Official Report from the Governor's Ad Hoc Commission of Enquiry, The Buffalo Creek Flood and Disaster, Charlston, W.V., 1972; and Bureau of Mines, Interim Report of Retaining Dam Failure No. 5 Preparation Plant, Buffalo Mining Company, Saunders, W.V., 26 February 1972.

Mine Safety and Health Administration, Final Report of Major Coal Mine Fire Disaster and Recovery Operations, Blacksville No. 1 Mine, Consolidation Coal Company, W.Va., 22 July 1972.

Bureau of Mines, Official Report of Major Mine Explosion Disaster, Itmann No. 3 Mine, Itmann Coal Company, Itmann, W. Va., 16 December 1972.

Mine Safety and Health Administration, Report of Investigation Underground Coal Mine Explosion, No. 2 Mine, P and P Coal Company, St. Charles, Va., 7 July 1977.

———, Report of Investigation Underground Coal Mine Inundation (Blackdamp), Moss No. 3 Portal A Mine, Clinchfield Coal Company, Duty, Va., 4 April 1978.

———, Report of Investigation Underground Coal Mine Explosion, Dutch Creek No. 1 Mine, Mid-Continent Resources Inc., Redstone, Colo., 15 April 1981.

UNITED KINGDOM

Her Majesty's Deputy Chief Inspector of Mines and Quarries, Report on the causes of, and circumstances attending, the explosion which occurred at Cardowan Colliery, Lanarkshire, 25 July 1960, HMSO, Cmnd. 1260.

H.M. Chief Inspector of Mines and Quarries, Report on the causes of, and circumstances attending, the explosion which occurred at Six Bells Colliery, Monmouthshire, 28 June 1960, HMSO, Cmnd. 1272.

H.M. Chief Inspector of Mines and Quarries, Report on the causes of, and circumstances attending, the explosion which occurred at Hampton Valley Colliery, Lancashire, 22 March 1962, HMSO, Cmnd. 1845.

H.M. Divisional Inspector of Mines and Quarries, Report on the causes of, and circumstances attending, the explosion which occurred at Tower Colliery, Glamorganshire, 12 April 1962, HMSO, Cmnd. 1850.

H.M. Chief Inspector of Mines and Quarries, Report on the causes of, and circumstances attending, the explosion which occurred at Cambrian Colliery, Glamorgan, 17 May 1965, HMSO, Cmnd. 2813.

Report of the Tribunal Appointed to Inquire into the Disaster at Aberfan, 21 October 1966, HMSO, H.L. 316.

H.M. Chief Inspector of Mines and Quarries, Report on the causes of, and circumstances attending, the fire which occurred at Michael Colliery, Fife, 9 September 1967, HMSO, Cmnd. 3657.

H.M. Divisional Inspector of Mines and Quarries, Report on the causes of, and circumstances attending, the outburst of coal and firedamp which occurred at Cynheidre/Pentremawr Colliery, Carmarthenshire, 6 April 1971, HMSO, Cmnd. 4804.

H.M. Chief Inspector of Mines and Quarries, Report on the causes of, and circumstances attending, the inrush which occurred at Lofthouse Colliery, Yorkshire, 21 March 1973, HMSO, Cmnd. 5419.

H.M. Chief Inspector of Mines and Quarries, Report on the causes of, and circumstances attending, the extensive fall of roof which occurred at Seafield Colliery, Fife, 10 May 1973, HMSO, Cmnd. 5485.

H.M. Chief Inspector of Mines and Quarries, Report on the causes of, and circumstances attending, the overwind which occurred at Markham Colliery, Derbyshire, 30 July 1973, HMSO, Cmnd. 5557.

H.M. Chief Inspector of Mines and Quarries, Report on the causes of, and circumstances attending, the explosion which occurred at Houghton Main Colliery, South Yorkshire, 12 June 1975, HMSO.

H.M. Inspectorate of Mines and Quarries, Report on the causes of, and circumstances attending, the locomotive manriding accident which occurred at Bentley Colliery, South Yorkshire, 21 November 1978, HMSO.

H.M. Inspectorate of Mines and Quarries, Report on the causes of, and circumstances attending, the ignition and explosion of firedamp which occurred at Golborne Colliery, Greater Manchester County, 18 March 1979, HMSO.

AUSTRALIA

Judge A.J. Goran, Report in the matter of an inquiry in pursuance of the Coal Mines Regulation Act into an accident which occurred at the Bulli Colliery, 9 November 1965, Sydney.

Mining Warden K. Hall, Accident at Box Flat Colliery, 31 July 1972, Brisbane.

Mining Warden E. N. Loane, Accident at Kianga No. 1 Underground Mine, 20 September 1975, Brisbane.

Judge, A. J. Goran, Explosion at Appin Colliery, 24 July 1979, Sydney.

Belgium

Rapport de la Commission d'enquête chargée de rechercher les causes de la catastrophe survenue au Charbonage du Bois-de-Cazier, le 8 août 1956, Ministry of Economic Affairs, Brussels.

Zimbabwe

Report of the Commission of Inquiry into the Wankie Colliery Disaster and General Safety in Coal Mines in Rhodesia, 1973, Salisbury.

# References

Bacow, Lawrence S. 1980. *Bargaining for Job Safety and Health.* Cambridge, Mass.: M.I.T. Press.

Bardach, Eugene, and Kagan, Robert A. 1982. *Going by the Book: The Problem of Regulatory Unreasonableness.* Philadelphia: Temple University Press.

Becker, Howard S. 1963. *Outsiders: Studies in the Sociology of Deviance.* London: Collier-Macmillan.

Bituminous Coal Operators' Association. 1977. *Federal Coal Mine Health and Safety Act of 1969: A Constructive Analysis with Recommendations for Improvements.* Washington, D.C.

Boden, Leslie I. 1983. "Government Regulation of Occupational Safety: Underground Coal Mine Accidents 1973-1975." Unpublished manuscript. Boston: Harvard School of Public Health.

Boden, Leslie I.; Zimmerman, Martin B.; and Spiegelman, Donna. 1981. *The Effects of Mine Safety and Health Administration (MSHA) Enforcement on the Cost of Underground Coal Mining.* Springfield, Va.: National Technical Information Service.

Braithwaite, John. 1980. "Inegalitarian Consequences of Egalitarian Reforms to Control Corporate Crime." *Temple Law Quarterly* 53: 1127-46.

———. 1982. "Challenging Just Deserts: Punishing White Collar Criminals. *Journal of Criminal Law and Criminology* 73: 723-63.

———. 1982b. "Enforced Self-Regulation: A New Strategy for Corporate Crime Control." *Michigan Law Review* 80: 1466-1507.

———. 1984. *Corporate Crime in the Pharmaceutical Industry.* London: Routledge & Kegan Paul.

Braithwaite, John, and Fisse, Brent. 1984. "Varieties of Responsibility and Organizational Crime," Unpublished manuscript. Canberra: Australian National University.

Braithwaite, John and Geis, Gilbert. 1982. "On Theory and Action for Corporate Crime Control." *Crime and Delinquency* April: 292-314.

Breslin, J. J., and Anderson, R. J. 1976. *Observations on Current American, British, and West German Underground Coal-Mining Practices.* Columbus, Ohio: Battelle Memorial Institute.

Bryan, Andrew. 1975. *The Evolution of Health and Safety in Mines.* Hertfordshire: Ashire Publishing.

Carson, W. 1981. *The Other Price of Britain's Oil: Safety and Control in the North Sea,* Oxford: Martin Robertson.

Chambliss, William J. 1967. "Types of Deviance and the Effectiveness of Legal Sanctions." *Wisconsin Law Review* Summer: 250-79.

Clinard, Marshall B., and Meier, Robert F. 1979. *Sociology of Deviant Behavior.* New York: Holt, Rinehart & Winston.

Clinard, Marshall B., and Yeager, Peter C. 1980. *Corporate Crime.* New York: Free Press.

Cocozza, Joseph, and Steadman, Henry J. 1978. "Prediction in Psychiatry: An Example of Misplaced Confidence in Experts." *Social Problems* 25: 265-76.

Coffee, John Collins, Jr. 1981. "No Soul to Damn, No Body to Kick: An Unscandalized Essay on the Problem of Corporate Punishment," *Michigan Law Review* 79: 413-24.

Cohen, Murray L.; Groth, A. Nicholas; and Siegel, Richard. 1978. "The Clinical Prediction of Dangerousness." *Crime and Delinquency* 24: 28-39.

Collinson, J. L. 1976. *Managing Health and Safety in a Period of Change.* Paper read at Symposium on Health, Safety and Progress, 27-29 October, at Harrogate, England.

_____. 1978. "Safety: Pleas and Prophylactics." *Mining Engineer* July: 73-83.

_____. 1979. "Safety-Risk Rationalization." *Mining Engineer* November: 411-7.

_____. 1980. "Safety—The Cost of Accidents and Their Prevention." *Mining Engineer* January: 561-7.

Conrad, John P., and Dinitz, Simon, eds. 1977. *In Fear of Each Other: Studies of Dangerousness in America.* Lexington, Mass.: Lexington Books.

Consolidation Coal Company. 1980. *Cost-Benefit Analysis of Deep Mine Federal Safety Legislation and Enforcement.* Pittsburgh, Pa.

Criminal Law and Penal Methods Reform Committee of South Australia. 1977. *Fourth Report: The Substantive Criminal Law.* Adelaide: South Australian Government Printer.

Dardalhon, Andre. 1964. "Evolution de la Securité dans les Mines: Minières et Carrières de 1841 à 1962." *Annales des Mines* 27-66.

David, John Peter. 1972. "Earnings, Health, Safety and Welfare of Bituminous Coal Miners Since the Encouragement of Mechanization by the United Mine Workers of America." Unpublished Ph.D. dissertation, West Virginia University.

Davis, Robert T., and Stahl, R. W. 1967. *Safety Organization and Activities of Award-Winning Companies in the Coal Mining Industry.* Bureau of Mines Information Circular 8224. Washington, D.C.

DeMichiei, John M.; Langton, John F.; Bullock, Kenneth A; and Wiles, Terrance C. 1982. *Factors Associated with Disabling Injuries in Underground Coal Mines.* Mine Safety and Health Administration. Washington, D.C.

Diver, Colin S. 1979. "The Assessment and Mitigation of Civil Money Penalties by Federal Administrative Agencies." *Columbia Law Review* 79:1435-502.

———. 1980. "A Theory of Regulatory Enforcement." *Public Policy* 28: 257-99.

Enterline, Philip E. 1964. "Mortality Rates Among Coal Miners." *American Journal of Public Health* 54: 761.

Ermann, M. David, and Richard Lundman. 1982. *Corporate Deviance.* New York: Holt, Rinehart & Winston.

Fisse, Brent. 1981. "Community Service as a Sanction Against Corporations." *Wisconsin Law Review* 1981: 970-1017.

Fisse, Brent, and Braithwaite, John. 1983. *The Impact of Publicity on Corporate Offenders.* Albany: State University of New York Press.

Fisse, Brent, and Braithwaite, John. 1984. "Sanctions Against Corporations: Dissolving the Monopoly of Fines." In *Business Regulation in Australia*, ed. Roman Tomasic. Sydney: CCH.

Franklin, Ben A. 1969. "The Scandal of Death and Injury in the Mines." *New York Times Magazine* 30 March: 25.

Friedmann, Wolfgang. 1972. *Law in a Changing Society.* 2nd ed. Harmondsworth: Penguin.

Geerken, Michael R., and Gove, Walter R. 1975. "Deterrence: Some Theoretical Considerations." *Law and Society Review* 9:509.

General Accounting Office. 1971. *Problems in Implementation of the Federal Coal Mine Health and Safety Act of 1969.* Washington, D.C.

General Accounting Office. 1981. *Low Productivity in American Coal Mining: Causes and Cures.* Washington, D.C.

Goldschmid, Harvey J. 1972. *An Evaluation of the Present and Potential Use of Civil Money Penalties as a Sanction by Federal Adminstrative Agencies.* Report in Support of Recommendation 72-6, Adminstrative Conference of the United States, pp. 896-964.

Gordon, R. L.; Manula, C. B.; Fettig, C., and Gresham, J. B. 1979. *Simulating the Effects of the Coal Mine Health and Safety Act.* University Park: Pennsylvania State University.

Greenspun, Julian S. 1982. "Criminal Intent Requirements and Defenses in Regulatory Prosecutions." *Criminal Law Journal* 6: 293-308.

Griffiths, John. 1970. "Ideology in Criminal Procedure *or* A Third 'Model' of the Criminal Process." *Yale Law Journal* 79: 359-417.

Grimaldi, John V., and Simonds, Rollin H. 1975. *Safety Management.* 3rd ed. Homewood, Ill.: Richard D. Irwin.

Gross, Edward. 1978. "Organizations as Criminal Actors." In *Two Faces of Deviance: Crimes of the Powerless and Powerful,* ed. P. R. Wilson and J. Braithwaite. Brisbane: University of Queensland Press.

Heaviside, Michael. 1980. *Legislative and Adminstrative Contents of Food and Drug Administration Data.* Washington, D.C.: Bureau of Social Science Research.

Henderson, David R. 1976 "The Economics of Safety Legislation in Underground Coal Mining." Ph.D. dissertation, Department of Economics, University of California, Los Angeles.

Hill, Forrest E.; Cook, Frank X.; Herhal, Albert J.; Krohta, Barbara J.; and Manual, Ernest H. 1981. *Analysis of the Labor Productivity Decline in the U.S. Bituminous Coal Industry.* Prepared by Emory Ayers Associates for the U.S. Department of Energy. New York.

Hopkins, Andrew. 1978. "Anatomy of Corporate Crime." In *Two Faces of Deviance: Crimes of the Powerless and Powerful,* eds. P.R. Wilson and J. Braithwaite. Brisbane: University of Queensland Press.

_____. 1984. "Blood Money? The Effect of Bonus Pay on Safety in Coal Mines," *Australian and New Zealand Journal of Sociology* 20: 23-46.

Kelman, Steven. 1981. *Regulating America, Regulating Sweden: A Comparative Study of Occupational Safety and Health Policy.* Cambridge, Mass.: M.I.T. Press.

Lewis-Beck, Michael S., and Alford, John R. 1980. "Can Government Regulate Safety: The Coal Mine Example." *American Political Science Review* 74: 745-56.

Lilly, Peter B. 1979. "Improving Productivity Through Effective Maintenance Management." In *How to Improve Productivity in Underground Coal Mining.* New York: Emory Ayers Associates.

Lipton, Douglas; Martinson, Robert; and Wilks, Judith. 1975. *The Effectiveness of Correctional Treatment: A Survey of Treatment Evaluation Studies.* New York: Praeger.

McAdams, John B. 1977. "The Appropriate Sanctions for Corporate Criminal Liability: An Eclectic Alternative." *Cincinnati Law Review* 46: 989-1000.

McAteer, J. Davitt. 1981. "Accidents: Causation and Responsibility in Law, a Focus on Coal Mining." *West Virginia Law Review* 83: 921-43.

McAteer, J. Davitt, and Galloway, L. Thomas. 1980. "A Comparative Study of Miners' Training and Supervisory Certification in the Coal Mines of Great Britain, The Federal Republic of Germany, Poland, Romania, France, Australia and the United States: The Case for Federal Certification of Supervisors and Increased Training of Miners." *West Virginia Law Review* 82: 937-1018.

Meisenhelder, Thomas. 1982. "Becoming Normal: Certification as a Stage in Exiting from Crime." *Deviant Behavior* 3: 137-53.

Mendeloff, John. 1979. *Regulating Safety: An Economic and Political Analysis of Occupational Safety and Health Policy*. Cambridge, Mass.: M.I.T. Press.

Merton, Robert K. 1968. *Social Theory and Social Structure*. New York: Free Press.

Mine Enforcement and Safety Administration. 1977. *A Report on Civil Penalty Effectiveness*. Washington, D.C.

Nagel, Trevor W. 1979. "The Fine as a Sanction Against Corporations.' Honours dissertation. Law School, University of Adelaide.

National Academy of Sciences. 1969. *Mineral Science and Technology*. Washington, D.C.

National Academy of Sciences. 1982. *Toward Safer Underground Coal Mines*. Washington, D.C.

National Research Council Comment on Underground Coal Mine Safety. 1982. *Toward Safer Underground Coal Mines* Washington, D.C.: National Academy Press.

Nierenberg, Gerard I. 1981. *The Art of Negotiating*. New York: Simon & Schuster.

Oak Ridge Associated Universities. 1979. *Determinants of Coal Mine Labor Productivity Change*. Washington, D.C.: Department of Energy and Department of Labor.

Orland, Leonard. 1980. "Reflections on Corporate Crime." *American Criminal Law Review* 17: 501-17.

Parnell, Nina. 1981. "The Coal Mines Regulation Act, N.S.W. From Instrumental to Symbolic Legislation: A Case Study in the Sociology of Law." Honours dissertation, Department of Sociology, Australian National University, Canberra.

Perry, Charles S. 1981. "Safety Laws and Spending Save Lives: An Analysis of Coal Mine Fatality Rates 1930-1979." Unpublished manuscript, Department of Sociology, University of Kentucky.

———. 1981. "Dying to Dig Coal: Fatalities in Deep and Surface Coal

Mining in Appalacian States, 1930-1978." Unpublished manuscript, Department of Sociology, University of Kentucky.

Pfeifer, C. Michael, Jr.; Stefanski, Joseph L.; and Grether, Craig B. 1976. *Psychological, Behavioral, and Organizational Factors Affecting Coal Miner Safety and Health.* Columbia, Md.: Westinghouse Behavioral Services Center.

President's Commission on Coal. 1980. *The Acceptable Replacement of Imported Oil with Coal.* Washington, D.C.: Government Printing Office.

Posner, Richard A. 1977. *Economic Analysis of Law.* 2d ed. Boston: Little, Brown and Co.

Rhodes, Gerald. 1981. *Inspectorates in British Government.* London: Allen & Unwin.

*Royal Commission on Safety in Coal Mines.* 1938. London: Her Majesty's Stationery Office.

Ruttenberg, Ruth. 1981. "Regulation is the Mother of Invention." *Working Papers for a New Society* 8: 42-47.

Saunders, Mark S.; Patterson, Terry V.; and Peay, James M. 1976. *The Effect of Organizational Climate and Policy on Coal Mine Safety.* Pittsburgh: U.S. Bureau of Mines Safety Research Center.

Schelling, Thomas C. 1974. "Command and Control." In *Social Responsibility and the Business Predicament*, ed. James W. McKie. Washington, D.C.: Brookings Institution.

Schrag, Philip G. 1971. "On Her Majesty's Secret Service: Protecting the Consumer in New York City." *Yale Law Journal* 80: 1529-603.

Shover, Neal; Clelland, Donald A.; and Lynxwiler, John. 1982. "Constructing a Regulatory Bureaucracy: The Office of Surface Mining Reclamation and Enforcement." Washington D.C.: National Institute of Justice.

Smith, R. 1974. "The Feasibility of an 'Injury Tax' Approach to Occupational Health and Safety," *Law and Contemporary Problems* 38: 730-44.

Solomon, Lewis D., and Nowak, Nancy Stein. 1980. "Managerial Restructuring: Prospects for a New Regulatory Tool." *Notre Dame Lawyer* 56: 120-40.

Stern, Gerald M. 1976. *The Buffalo Creek Disaster: The Story of the Survivors' Unprecedented Lawsuit.* New York: Random House.

Stone, Christopher D. 1975. *Where the Law Ends: The Social Control of Corporate Behavior.* New York: Harper Torchbooks.

Subcommittee on Labor of the Committee on Human Resources, U.S. Senate. 1978. "Legislative History of the Federal Mine Safety and Health Act of 1977." 95th Cong., 2d sess. Washington, D.C.

Sutton, Adam, and Wild, Ron. 1978. "Corporate Crime and Social Structure." In *Two Faces of Deviance: Crimes of the Powerless and Powerful,* ed. P. R. Wilson and J. Braithwaite. Brisbane: University of Queensland Press.

Thompson, Anthony J., and Sliter, Charles E. 1981. Planning for federal safety and health inspections. In *Mine Safety and Health Litigation,* American Mining Congress, Washington, D.C.

Turton, F. B. 1981. "Colliery Explosions and Fires: Their Influence upon Legislation and Mining Practice." *Mining Engineer* September: 157-64.

Van Dine, Stephen; Conrad, John P.; and Dinitz, Simon. 1979. *Restraining the Wicked.* Lexington, Mass.: Lexington Books.

Veljanovski, Cento G. 1983. "Regulatory Enforcement: An Economic Study of the British Factory Inspectorate." *Law and Policy Quarterly* 5: 75-96.

Viscusi, W. Kip. 1983. *Risk by Choice: Regulating Health and Safety in the Workplace.* Cambridge, Mass.: Harvard University Press.

Waldman, Don E. 1978. *Antitrust Action and Market Structure.* Lexington, Mass.: Lexington Books.

West, D. J., and Farrington, D. P. 1977. *The Delinquent Way of Life.* London: Heinemann.

Yale Law Journal. 1979. "Structural Crime and Institutional Rehabilitation: A New Approach to Corporate Sentencing." *Yale Law Journal* 89: 353-75.

Yoder, Stephen A. 1978. "Criminal Sanctions for Corporate Illegality." *Journal of Criminal Law and Criminology* 69: 40-58.

Zabetakis, Michael G. 1981. "Productivity and Safety in U.S. Bituminous Coal Mines." *Mining Congress Journal* June: 19-21, 45.

Zelonka, J. Richard. 1974. *HSAC Technical Progress Report No. 4: Coal Mining Injuries and Noncompliance with Safety Regulations.* Denver: Mining Enforcement and Safety Administration.

Zimmerman, Martin B. 1981. *The U.S. Coal Industry: The Economics of Policy Choice.* Cambridge, Mass.: M.I.T. Press.

# Index